Material Culture of Breweries

GUIDES TO HISTORICAL ARTIFACTS

Books in this series are comprehensive guides to classes of historical artifacts commonly found in excavations, archives, museums, and private collections in North America and across the globe. Designed for archaeologists, material culture specialists, museum professionals, historians, decorative arts scholars, and collectors, the books explore the interconnections between objects and social identity, and are amply illustrated to aid in the identification and interpretation of the technical, temporal, and diagnostic significance of objects described. The series also serves to collate disparate and often obscure specialist literatures and provides an arena for contributors to explore the role of individual objects or assemblages in social action within communities.

Series Editors

Carolyn L. White is Assistant Professor of Historical Archaeology at the University of Nevada, Reno. She is particularly interested in marginal artifact types and the use of material culture to understand identity construction.

Timothy James Scarlett is Associate Professor of Historical Archaeology in Michigan Technological University's Industrial Heritage and Archaeology program. His research interests focus on material culture's role in the interplay of systems of cognition, belief, and economy.

Books in the Series

Volume 1, *Chinese Export Porcelains*, Andrew D. Madsen
Volume 2, *Material Culture of Breweries*, Herman Wiley Ronnenberg
Volume 3, *Ceramic Makers' Marks*, Erica Gibson

Material Culture of Breweries

Herman Wiley Ronnenberg

Walnut Creek, CA

Left Coast Press, Inc.
1630 North Main Street, #400
Walnut Creek, CA 94596
www.LCoastPress.com

ISBN 978-1-59874-166-7 hardcover
ISBN 978-1-59874-167-4 paperback

Library of Congress Cataloging-in-Publication Data
Material culture of breweries/Herman Wiley Ronnenberg
p. cm. -- (Guides to historical artifacts)
Includes bibliographic references and index.
ISBN: 9781598741667 (hardback : alk. paper)
ISBN: 9781598741674 (paperback : alk. paper)
1. Breweries --United States --History. 2. Breweries --United States --Equipment and supplies --History. 3. Industrial archaeology --United States. I. Ronnenberg, Herman Wiley.

TP573.U6 R764 2009 663/.420973 2009019103

Printed in the United States of America

The paper used in this publication meets the minimum requirements of American National Standard for Information Sciences—Permanence of Paper for Printed Library Materials, ANSI/NISO Z39.48–1992.

Book design and production by Lisa Devenish, Devenish Design
Cover design by Piper Wallis

Contents

DEDICATION

I would like to dedicate this book to the three men I worked for and with on large historical archaeology projects in Virginia and Idaho: the late James Deetz, Darby Stapp, and especially Roderick Sprague, whose suggestion for a book on breweries led directly to the writing and publication of this volume. Digging in the dirt is rewarding—it offers the thrill of discovery, expansion of thinking, and the joy that comes from working with people of ideas and action, such as these men.

LIST OF ILLUSTRATIONS

APPENDIX IV

ACKNOWLEDGMENTS

My deepest appreciation goes to Dr. Roderick Sprague who told me of the series of books on archaeology being undertaken by Left Coast Press and suggested I propose a book on brewing. I took classes from Dr. Sprague over twenty-five years ago and participated in archaeological work that he supervised in Warren, Idaho. He served on my doctoral committee. This project would never have happened without his initial suggestion.

Mia T. Parson, Supervisory Archaeologist/Section 106 Coordinator, Harpers Ferry National Historical Park, West Virginia, provided information on research done on the breweries of Harpers Ferry. Rachel Feit of Hicks and Company, Austin, Texas provided their reports and photos on the Schneider beer vault study. Marlin R. Ingalls provided a copy of his report on the Law Brewery Cave of Dubuque County, Iowa. Carl Adam, malting chemist for Anheuser-Busch at Idaho Falls, gave suggestions, corrections, and material on malting. Todd McMakin of the Texas Parks and Wildlife Department, sent me the report and photographs on the Kreische Brewery. Dr. John A. Walthall, Dr. Thomas E. Emerson, and Stephanie Daniels were kind enough to provide a copy of Floyd Mansberger's Illinois Department of Transportation report on the Kun-Rudolph Brewery of Springfield. Marlin R. Ingalls provided a copy of Leah D. Roger's report on the City Brewery of Des Moines, Iowa. My editor, Tim Scarlett, added a great deal to the organization and conceptualization of this book, and helped to smooth my convoluted use of the English language. Steve Seidenberg's editing to tidy up the nearly finished manuscript was most valuable and highly appreciated. He showed remarkable skill and patience with me. I hope I have not overlooked anyone. Without you fine people, my efforts would have been anemic indeed.

INTRODUCTION

The history of the American brewing industry both highlights and reflects much of the social, political, agricultural, technological, and ethnic history of the nation. This book outlines the evolution of the brewing industry in America with an emphasis on the physical remains of beer making in architecture, equipment, and packaging. In this introduction, I review the historiography of brewing and summarize the research questions that scholars have identified as interesting areas for future work. I include my own research in this discussion.

In the chapters that follow, I provide tools with which researchers can undertake their own studies of brewing processes, architectural evolution, and packaging materials. I emphasize useful information from makers' marks, patents, and other equipment that a researcher can use to generate chronological and typological information for a study of brewing or of a particular brewery. I supplement the main text with three appendices. These cover a chronological history of brewing in the Americas, with a focus on the United States; specialized terms used in this book; and organizations for collectors of brewing artifacts—*breweriana*, a term coined to describe items made by or for breweries.

In the years since 1980, increased interest in local beer and breweries has coincided with or fueled increased interest in the study of brewing history. Many excellent national, state, regional, and city histories of brewing and breweries now exist, as well as biographies of individual brewers, yet only two scholarly studies of single brewing companies exist: Cochran's *The Pabst Brewing Company: The History of An American Business* (1948) and Plavchan's *A History of Anheuser-Busch, 1852–1933* (1969). Cochran was almost apologetic for undertaking his study. I believe there are still attitudes and ideas left over from Prohibition that intimidate scholars and stifle study of the entire subject. Studies of breweries are seldom taken seriously. Some information on the industry is readily available through standard sources, and more appears all the time, but much about these businesses has never been studied or understood, and erroneous clichés still substitute for serious knowledge. In 1976 Diane Newell offered a bleak prophecy: "Many breweries will go simply because they are not taken seriously and they are not taken seriously because, to a large extent, their history and importance are not known or understood."[1] In the three decades since her prediction, some breweries have been saved, some have not, but many entirely new breweries have come and gone without benefit of analysis and without being adequately recorded.

A veritable army of breweriana collectors, associated in four major collector clubs and many smaller ones, strive to preserve interest in brewing history and collectibles. These collectors buy, sell, categorize, and trade their treasures. Serious breweriana collectors research the origins of their materials, photograph them, and often write about them in club newsletters and magazines, blogs, and websites. They do not produce peer-reviewed studies, and the quality of this information varies, but the voluminous amount and unique quality of the work make these vital sources for identification.

Collectors own massive numbers of bottles, cans, trays, labels, closures, openers, kegs, photographs, and dozens of other artifact types that can aid scholars trying to make comparisons and that tell a story of these manufacturing industries. Between them, the Breweriana Collectibles Club of America (BCCA, formerly the Beer Can Collectors of America), National Association of Breweriana Advertising (NABA), American Breweriana Association (ABA), and Eastern Coast Breweriana Association (ECBA), all of which are described in Appendix III, have thousands of members who live and breathe breweriana.

In 1976, Manfred Friedrich and Donald Bull published *The Register of United States Breweries, 1876–1976*, in two volumes. This was the first attempt to list all American breweries in a single source. In 1984, the same authors worked with Robert Gottschalk to update this information in *American Breweries*. The latest effort to publish such a list, *American Breweries II* by Dale P. Van Wieren, was published in 1995.[2]

Despite these monumental projects, a few breweries were not listed. For example, although breweries needed a federal permit after 1862, some breweries never bothered with permits or were in operation only briefly and never entered the federal records. Breweries before the 1860s are even more difficult to locate, since no single source exists from which to begin researching brewery names and locations.

A researcher must understand the overall historical milieu in which breweries operated to properly interpret archaeological finds. For instance, to understand the piping of beer from brew house to bottling house, the investigator must know the history of the United States Internal Revenue regulations for taxing beer. Brewers did not pipe beer to bottling houses for years because the law required the beer to be barreled and transported over a public road before bottling. Without this sort of knowledge, sound interpretation would be impossible. Likewise, a basic understanding of laws that regulated beer strength, license fees, and so forth helps interpret findings realistically.

The number of inventions and developments in brewing equipment, as well as the quantity of patent numbers, is monumental. Manufacture of refrigeration devices alone involved a dozen companies in the late nineteenth century and many dozens of patents. Obsolete brewing equipment generally was moved out of closed breweries, sold to other breweries—sometimes half a world away—or scrapped for the considerable metal involved. Only a minimal amount of equipment may be expected to remain in an excavation site. One exception to this that I have personally examined is the Gilbert Brewery of Virginia City, Montana, which looked in 1995 as though the last brewer was merely on vacation and expected to return and start his next brew immediately. It closed in 1935 but is still used as a saloon.

Most of the specific examples I discuss in this book come from my own research, which first centered on the breweries of Idaho and later spread to surrounding states. Except for a giant shipping brewery, this area has examples of most brewery types after 1855.

For studying brewing equipment and structures of the nineteenth century, there is no single source equal to the great compendium authored by the anonymous staff of *The Western Brewer* entitled *100 Years of Brewing*. It is physically the largest book on American brewing history and was nearly encyclopedic in its coverage up to its date of publication in 1903. Another excellent reference is the 1891 Charles Zoller Brewers Catalog (reprinted in the readily available *Richmond Beers*) had illustrations of all the equipment Zoller sold and could be invaluable for identifying equipment and supplies. Susan Appel's dissertation discusses American brewery architecture in a most authoritative way; she covers Cincinnati and St. Louis and touches on other cities, and has since written about Chicago. Many other brewing centers, such as Milwaukee, Philadelphia, and New York could provide the locale for similar architectural studies.

After the lager revolution, breweries present an avenue to study the influence of Germans on American culture, and a number of historians have worked on this issue. American breweries were also owned by the Dutch, English, Czechs, Poles, Irish, and Jews. An archaeological study of any example of these breweries could contribute important knowledge to this neglected aspect of brewing history.

The enormity of the subject of brewing on the North American continent may not be obvious at first, but becomes clear when we consider just a few possibilities. Dutch breweries stood in the area of modern New York in the 1600s. English breweries were in place, or at least brewing was being done, in early Virginia, later along the whole east coast, and by the mid-1800s into the Mississippi Valley. Breweries were established by the colonial French from Kaskaskia, Illinois, to Canada. Pre-lager German breweries may have been located where small colonies of Germans settled. Breweries of German origin or ancestry were all over the West from the 1840s on. Dozens of tiny mining camps in the Rocky Mountain region had breweries, though the size, age, technology, and architecture of these breweries all differed markedly. The archaeologist, therefore, must be flexible.

With the remodeling of so many obsolete breweries of the late nineteenth and early twentieth centuries, many future archaeological projects may involve nondestructive analysis of these structures. For instance, three architectural theses at the University of Idaho addressed remodeling plans for the defunct Schade Brewery in Spokane, Washington.

Many extant breweries are available to study if one needs standing structural sites for comparison. For instance, Bob Skilnik listed seventeen old Chicago breweries still standing as of 1999.[3] Jerry Apps, noting "Wisconsin is virtually a brewery graveyard," listed seventy breweries still standing in that state.[4] Even Idaho has two pre-Prohibition breweries still standing.

Basic Questions Addressed by the Archaeology of Breweries

Maureen Ogle, in her 2006 book, *Ambitious Brew: The Story of American Beer*, posed some very basic questions and dilemmas that she encountered, and with which any researcher, historian, or archaeologist of the American brewing industry must wrestle.[5] First, she asked:

> Did the world really need a rehash of this apparently well-known story? . . . I discovered that almost every aspect of that oft-told tale of skullduggery, greed and woe was false and that the truth was considerably more interesting and complex. Beer's history . . . embodies the essence of what it is to be American: our ambivalent attitude toward alcohol, our passion for invention and creativity, and our seemingly limitless ability to take

> old ideas and things and remake them in our uniquely American image.

Ogle used documents to reach her conclusions, but material culture studies have the potential to add even more to illuminate the story of brewing in America.

In his 1998 dissertation, Martin Stack took aim at much of the accepted historical story of American brewing. His work raised many questions that call for improved research models that include material culture. Also, for archaeologists to understand and interpret their finds, they need to understand the history as well as the current state of American brewing, as derived from written records. Unfortunately, some of that history is incomplete. For example, when Fred Mansberger began to investigate the 1850s Kun Brewery of Springfield, Illinois, he found that:

> . . . the early to mid-nineteenth century brewery industry in Illinois and the Midwest is poorly documented. Except for a handful of lithographs depicting mid-nineteenth century vernacular brewery buildings, little exists documenting the size and spatial layout of the mid-century brewery complex.[6]

This period of brewery complexes remains a wide open avenue of investigation.

Stack's dissertation examined the American brewing industry from 1865 to 1940. What he discovered contrasted with the accepted outline, leading him to propose several questions in regard to brewing history. These questions are ripe for archaeological and material culture research. For instance, historians presume that mass production is organized efficiently and economically, and can therefore put lower priced goods on the market. Is mass production's efficiency in its technology, organization, or in its ability to economize on transaction costs? Is the mass-production technology more efficient than possible small-scale alternatives, and are the products of mass production therefore of higher quality?

Perhaps inefficient technologies become locked in by tradition, rather than undergoing a survival-of-the-fittest test. Did a host of cultural, social, and organizational factors interact in shaping the technological path of the brewing industry? At the production level, are brewing companies active at introducing new technologies and processes? In postproduction, brewing companies actively market changes in packaging to get consumer acceptance. During pivotal decades consumers were apparently persuaded to prefer the packaged, pasteurized, shipped beer that dominated the market.[7] Was the ascendancy of bottled beer a natural response to changing consumer preferences, or the result of effective advertising? Likewise, historians explain the ascendancy of lager beer over ale by looking at immigration patterns, but may not have the whole story. Were changes in the brewing process and resultant beer styles really driven by consumer preferences, or did the change in preference among consumers from heavy British ales to lighter German lagers relate to the increase in available food, and the consequent decrease in the need for calories in beverages?

Other authors have examined other aspects of American brewing history and have posed different questions. Susan Appel's dissertation raised numerous questions about brewery architecture development; many of these queries could be enlightened by archaeological data. For example, the movement of midwestern breweries from primitive to sophisticated has been poorly recorded. Archaeology could aid in understanding that era of development.[8] Further, what led to the popularity of Romanesque architecture for breweries after the 1860s? And what led to its falling out of favor by the second decade of the twentieth century? Since brewery architecture development had a ten to fifteen year hiatus in America during Prohibition, from where did the design concepts for the brewery structures from 1930 to 1940 come?

David V. Burley and Michael Will studied the Klondike Brewery to understand larger issues of economic and political life in the Yukon during the era in which it operated. They hoped as well to discover how the brewery fit into the larger picture of a brewing industry "enmeshed in a dramatic state of technological change."[9].

The research undertaken by Leah Rogers for the excavation of the City Brewery of Des Moines, Iowa, sought to understand where and when each building was constructed, and how the brewers used each in the many activities at the complex. A question relating to buildings asked if any industrial activities were carried on outside, or was everything done inside? [10] Concerning consumer behavior patterns, Rogers asked the interesting questions: "Are these patterns reflective of the socioeconomic status of the brewery owner, the German ethnicity of the occupants, and/or the changing role of women in the nineteenth century?" Social history patterns relating to labor and management, and the disruptions caused by temperance legislation, may be indicated by the material culture.

I would pose additional questions that have been rarely addressed. An old craft—brewing—met industrialization in the nineteenth century. Material culture can contribute to understanding this phenomenon and add to established explanations. In addition, English- and German-based American breweries have been frequently studied, but smaller breweries operated by Americans of Polish, Czech, Irish, and other immigrant backgrounds have not. Archaeological studies could alter the prevalent ethnic generalizations and help illuminate, for example, whether ethnic differences in equipment, sources of supplies, techniques, and so on are noteworthy in the material remains. Researchers could compare these breweries with examples in the European countries from which their operators came.

For example, no above-ground examples of single-story hop-drying houses like the English oasts have been discovered in America, yet archaeologists could identify them if they ever existed. Similarly, little is known about how Asian or Scandinavian builders of the Pacific Northwest developed their ideas of hop house architecture and construction.[11] In time, archaeologists could study enough examples to develop an improved chronology.

Breweries have caused ecological damage by accidental spills or by deliberately flushing materials into streams. Have changes in architecture or equipment made accidents less likely? Breweries also frequently had financial involvement in saloon ownership; such "tied" saloons sold the beer of a single brewery exclusively. Did these saloons ever bottle beer, or maintain transportation items such as wagons and horses for their brewery partner, or assist the larger brewery business in other ways? Archaeological work at saloons affiliated with a particular brewery could address local questions about supplies, customs, transportation, and other aspects of the saloon/brewery relationship. Prohibition-era breweries could be analyzed to discover how they engaged in illegal activities without being caught. Piping, in particular, could be reconstructed from archaeological data.

Hermann Schlüter, scholar of American brewery labor, alleged in 1910 that brewery owners mistreated workers.[12] For example, managers would strike workers, force them to board at a brewer's home or boarding house, deliberately intoxicate them to produce a workforce of zombies, and so on.[13] Could exhausted men read steam gauges, saccharometers, and thermometers of normal size accurately; could they lift 100-pound blocks of ice by the end of a long shift? Were conditions for brewery workers better in areas where labor was scarce and valuable—such as the West—and how did treatment change over time? What was the relationship between the physical brewery building and the workers who spent 100 hours a week there?

Did the consumption of beer by brewery workers alter in a detectable way over time and particularly after breweries produced only near beer? When the nature of the beer changed radically during Prohibition, did the nature of the social relations change radically as well? Did the change in the material culture produce change in some unexpected way?

British scholars would define the part of this study up to 1750 as *Post-Medieval Archaeology*, and the later period as *Industrial Archaeology*.[14] The earliest breweries in English North America do appear to be continuations of the European tradition as practiced for the previous several hundred years. The development of brewing in America from a small craft to a giant industry is a case study in capitalism, and fits perfectly into the "archaeology of capitalism" theoretical school.[15]

Archaeologists will generally, but not always, be able to locate documentary sources about a given brewery sufficient to be a second line for evidentiary analysis. Documentary archaeologists may find "these additional windows may provide overlapping, conflicting, or entirely different insights into the past. The challenge for archaeologists is to use these independent but complementary lines of evidence to construct meaningful, fuller understandings of the past."[16] Very few breweries saved their business records and donated them to museums and universities, but a few did, and those contain a wealth of information on daily details.[17]

Laurie Wilkie found evidence from the Zeta Psi fraternity house that indicated that near beer replaced the use of real beer, and drinking customs at the house went on during Prohibition. She explained, "beer drinking remained an important expression of fictive kinship."[18] Did on-the-job beer consumption by brewery workers likewise create fictive kinship?

An account of brewing history produced by the historical archaeologist must consider that "in their encounters with the contradictions between words and things, historical archaeologists always live with the knowledge that there is no single story that can adequately account for the phenomena they study."[19] There may be a place in the brewing story to unearth "small things forgotten," particularly, in the daily activities of the workers. Very few brewery workers left accounts of their daily labors and events.

Historical archaeology is often used to study colonialism and the transfer of culture from old society to new. A study, for instance, comparing the brew house erected on William Penn's estate with similar contemporary (1863) estate brew houses in England could see if any adaptations from English tradition were made to accommodate the new location and circumstances. In a similar vein, do very early breweries reflect the organic, communal worldview that Deetz described, and is there a change when the Georgian worldview replaces it? Was the balanced symmetry of Georgian houses reflected in breweries constructed during the Georgian age?[20] If cooks in New England preferred to bake and those in the Chesapeake made puddings due to their regional origins in England, did any comparative difference in brewing style exist?

Both German and English cultures are known for hierarchical arrangement. In larger breweries the brewmaster was in charge of everyone and everything in the factory. Does this reflect a hierarchical worldview, and did the English and German views differ measurably? Did the office workers rate higher than production workers, and was this noticeable in the physical plant, such as in more opulent restrooms?

Studies of New York found that both Jewish and Irish immigrants drank more wine than beer in their homes. Does this pertain to the local brewing tradition? Breweries near Five Points, New York City, were confined to foul-smelling areas near tannery yards where both industries spewed byproducts into a nearby pond.[21] Archaeologists could analyze the landscapes surrounding breweries to see if this was a widespread practice. How close to universality was it? Is this strictly a large urban phenomenon?

In Support of the Champions of Brewery Research

Every historian I know who has done research on beer or brewing has faced a general response that his time could be better spent elsewhere. Perhaps as a holdover from the attitudes and prejudices of Prohibition, Americans somehow cannot comprehend the seriousness of the study of an industry that employs, directly or indirectly, hundreds of thousands of people; invests billions in capital; tries to comply with thousands of federal, state, county, and city ordinances; and has ubiquitous advertising and distribution. Few serious scholars have been willing to take the risks involved in brewery research—principally, the risk of not being taken seriously. Overall, additional archaeology of brewery sites could add immeasurably to our general knowledge of the evolution of brewing as well as the history of the individual brewery studied. Some archaeologists have studied breweries despite the general attitudes and made bold additions to our understanding of their material culture. There is much left to learn, and I hope this volume will assist those who will do that work in the future.

NOTES

1. Diane Newell, "With Respect to Breweries," *Historic Preservation* 27 (January 1976): 24.
2. See Manfred Friedrich and Donald Bull, *The Register of United States Breweries, 1876–1976*. 2 vols. (Trumbull, CT: Donald Bull, 1976). See also Donald Bull, Manfred Friedrich, and Robert Gottschalk, *American Breweries* (Trumbull, CT: Bullworks, 1984); Dale P. Van Wieren, *American Breweries II* (West Point, PA: Eastern Coast Breweriana Association, 1995).
3. Bob Skilnik, *The History of Beer and Brewing in Chicago, 1833–1978* (St. Paul: Pogo Press, 1999), 217–221.
4. Jerry Apps, *Breweries of Wisconsin* (Madison: University of Wisconsin Press, 1992), 174.
5. Maureen Ogle, *Ambitious Brew: The Story of American Beer* (Orlando: Harcourt, Inc., 2006), ix.
6. Floyd Mansberger, "The Industrial Archaeology of Breweries: Archaeological Investigations at the Mid-Nineteenth Century Kun-Rudolph Brewery, Springfield" (Springfield, Illinois: Fever River Research [for Illinois Department of Transportation], 1993), 1.
7. Martin Heidegger Stack, "Liquid Bread: An Examination of the American Brewing Industry, 1865–1940" (dissertation, University of Notre Dame, 1998), 2–6, 114.
8. Susan K. Bigley Appel, "The Midwestern Brewery Before Prohibition: Development of an American Industrial Building Type" (dissertation, University of Illinois, 1990).
9. David V. Burley and Michael Will, "Special Brew: Industrial Archaeology and History of the Klondike Brewery," *Occasional Papers in Archaeology* 11, *Hudé Hudän* Series (Heritage Branch: Government of the Yukon, 2002), 1.
10. Leah D. Rogers, " 'It Was Some Brewery': Data Recovery of the City Brewery Site, 13PK661, Des Moines, Iowa" (Des Moines, Iowa: Des Moines Metropolitan Transit Authority, 1996), 5.
11. Michael A. Tomlan, *Tinged with Gold: Hop Culture in the United States* (Athens: University of Georgia Press, 1992), 219.
12. Hermann Schlüter, *The Brewery Industry and the Brewery Workers' Movement in America* (1910; repr., New York: Burt Franklin, 1970). 89.
13. William L. Downard, *The Cincinnati Brewing Industry, A Social and Economic History* (Oberlin: Ohio University Press, 1973), 99.
14. Dan Hicks and Mary C. Beaudry, eds., *The Cambridge Companion to Historical Archaeology* (Cambridge: Cambridge University Press, 2006), 3.
15. The emphases in American historical archaeology have been cultural evolution, adaptation, cultural differentiation, shifting worldviews, and capitalism. See Hicks and Beaudry, eds., *The Cambridge Companion to Historical Archaeology* (Cambridge: Cambridge University Press, 2006), 5.
16. Laurie A. Wilkie, "Documentary Archaeology," in Hicks and Beaudry, eds., *The Cambridge Companion to Historical Archaeology* (Cambridge: Cambridge University Press, 2006), 14.
17. Several breweries from Spokane, Washington, merged, and when they later closed, many stacks of records were donated to Washington State University in Pullman, Washington. This is one of the larger such deposits.
18. Wilkie, "Documentary Archaeology," 28.
19. Rosemary Joyce, "Writing Historical Archaeology," in Hicks and Beaudry, eds., *The Cambridge Companion to Historical Archaeology* (Cambridge: Cambridge University Press, 2006), 48–49.
20. Susan Lawrence and Nick Shepherd, "Historical Archaeology and Colonialism," in Hicks and Beaudry, eds., *The Cambridge Companion to Historical Archaeology* (Cambridge: Cambridge University Press, 2006), 72.
21. Tadhg O'Keeffe and Rebecca Yamin, "Urban Historical Archaeology," in Hicks and Beaudry, eds., *The Cambridge Companion to Historical Archaeology* (Cambridge: Cambridge University Press, 2006), 97, 98.

CHAPTER ONE

Overview of Brewing Techniques With Emphasis on Changes in Equipment and Raw Materials: 1500–2000

Beer came to America in the ships of the first European explorers,[1] and was ubiquitous as a provision on ships. The idea of brewing beer came to America with the first settlers. European beer was imported across the Atlantic but the supply was spotty, and the quality generally deteriorated on the voyage. If settlers from beer drinking cultures, especially the British and the Dutch, wanted their accustomed beverages, they realized that they had to make their own. Manufacture required raw materials, equipment, and skill. In the New World, adjustments had to be made in all of these to create beer, or something resembling beer. Human ingenuity worked wonders, and the seed of the American brewing tradition was sown.

In the prescientific age, the brewing process was mysterious; this caused successful, quality brewing to be honored and even held in awe. Between the raw material and the final product lies microscopic chemistry so complex that for millennia, it was considered nearly miraculous.

No matter how primitive or sophisticated a system for brewing might be, the processes involved are the same. The raw materials are water, grain (usually barley), yeast, and hops or other flavoring agents. All of these must be obtained, stored, and processed. Boiling, kiln drying, and cooling are indispensable, so fuel and a technique for cooling are required. The final product—*beer*—consists of about 90 percent water, 4 percent alcohol, 6 percent dissolved sugars from the grain, and a tiny amount of hops extract for flavor.[2]

A brewer must also put the beer into vessels, transport the vessels to a retailer (if this is not the brewer himself), and the retailer must serve the beer to the consumer in some vessel. All of these items reflect the culture that produced them and the level of science and technology known and affordable at that time and place.

A physical facility to accomplish this work is also required, but it might be small and temporary, such as a shed or barn, or enormous and grandiose such as the major commercial breweries of the twenty-first century. Equipment for each part of the brewing process is necessary, and highly varied.

When the first English colonists at Roanoke tried malting maize to make beer, they had only hand labor and primitive tools with which to work. By the early twenty-first century brewing involved equipment and materials so specialized and complicated that universities offered degrees in the subject. The development of the apparatus of brewing left a physical record of the progress of the brewing industry. The recovery and analysis of the various items of brewing equipment will show the development of not only brewing techniques, but also scientific concepts and their application to industry of all types, trade patterns, and cultural change.

The Basic Recipe of Beer

European brewing has roots that stretch back into prehistory. Even in a prescientific age brewers learned how to make beer efficiently from practical applications. The process begins with preparing the grain. Barley was and is the dominant grain in brewing, but wheat, rye, oats, and others have all been used independently or mixed with barley. Raw grain is high in starch and low in sugar and will not ferment rapidly enough, if at all. The solution is to sprout—called malting—the grain to change much of the starch to sugar. The grain is moistened and the seeds begin to grow in a few days, just as they would if they had been planted. A sprout grows out the end and is allowed to develop until it is the length of the seed from which it emerged, and to which it is still attached. The grain at this stage is called malt and must be dried to prevent rotting. The malt may be dried with virtually no change in color or roasted to a darker color, reaching nearly black. This gives the final beer its color. After drying, the malt may be transported or stored.

When ready for use, the malt is ground much as flour would be, although not as finely. The ground malt is steeped in hot water for some hours to allow the sugars to dissolve into the water; this is called mashing. The residual grain, with a low starch and sugar content, is filtered out. The sugary liquid, now called wort (pronounced *wurt*) is boiled for several hours with added hops. The hops are filtered out when the boiling concludes. The wort is cooled to the proper temperature and yeast is then added in a process called pitching. Fermentation takes place as the yeast acts on the sugar. This process may take several days or several weeks depending on the temperature and type of yeast. The fermentation process changes the sugar to alcohol and carbonic gas.

The beer is often filtered, and then aged to reach maximum flavor. It is then transferred to containers to be transported to the wholesaler, retailer, and consumers. Wooden barrels were used for centuries, followed by galvanized barrels, then aluminum. For small quantities beer was put into glass or stoneware bottles. In the twentieth century beer was first put into metal cans.

Every step in the brewing process must be tightly controlled to get the desired final product. Improper brewing techniques can cause an entire batch of beer to be ruined, or result in poor quality.

The Beer Revolution of the Nineteenth Century

Starting in the 1840s and moving rapidly after 1860, the replacement of the British brewing tradition with the German brewing tradition changed America. Lager beer slowly supplanted ale, which may have been representative of much wider cultural change. In 1880, beer authority Frederick William Salem said: "when we remember that outside larger cities, even twenty years ago, ale was almost sure to be dull and muddy and very apt to be sour, we must admit that American ale-brewers have accomplished much."[3]

The latter half of the nineteenth century saw science and technology applied to brewing to such an accelerated extent that it was a virtual revolution. A master brewer of 1800 would have been awestruck and lost in a modern brewery of 1900. A master brewer of 1900 would certainly be impressed by a brewery of 2000, but with the exception of the various uses of computers, the manufacturing process at the end of the twentieth century would be familiar to a brewer of the late nineteenth century. Thousands of inventions in Europe and the United States—and thousands of protective patents—chronicle the advances. Europe and America cross-fertilized each other with brewing knowledge in the nineteenth century. Brewers often visited across the Atlantic specifically for research purposes. The extensive literature on European brewing history, biography, and archaeology would prove instructive for understanding the American experience and for making comparisons.

Early Beer Varieties

Over fifty distinct styles of beer are currently known in the Western world. During the exploration and colonial periods, British beer was categorized into three basic types. Small or single beer was a weak variety made only from malt. Middle, also known as ships or table beer, had some sugar added to the malt to increase the alcohol content. Double, old, or strong beer had enough sugar added to assure a high alcohol content.[4]

Each part of the brewing process must be understood in greater depth if the archaeologist is to identify and understand cultural and industrial items that are uncovered.[5]

Brewing Materials

As mentioned, beer requires several basic ingredients: water, grain, yeast, and flavorings. Even subtle variations in the quality or species of these elements can have a profound effect on the finished result. Through trial and error, brewers are able to refine their product into their ideal concept of the perfect beer.

Water

Many American breweries in the modern era have touted their local water as an almost secret ingredient in their beer. Olympia Brewing Company advertised with the slogan "It's the Water!" The artesian water of Tumwater, Washington, was responsible for its distinctive flavor. Coors of Golden, Colorado, promotes its Rocky Mountain spring water. Genesee Brewing of New York claimed that the best brewing water in the world came from nearby Hemlock Lake. The now-defunct Kessler Brewery of Helena, Montana, was located next to a gushing mountain spring; Herr Kessler received a land patent from President Grant assuring him possession of the famous Kessler Springs.[6] Leinenkugel of Chippewa Falls, Wisconsin has always used the water of the Big Eddy Springs. Hamm's Brewing, when based in Minneapolis, was "From the Land of Sky Blue Waters"; later, when the corporation had breweries in multiple locations, the beer was "Born in the Land of Sky Blue Waters."

A number of breweries ran sideline businesses selling bottled water. Cold Spring Brewing Company of Cold Spring, Minnesota, marketed mineral water when founded in 1874 and again decades later.[7] Carl Mallon, while building a new brewery at Wallace, Idaho, in 1890, discovered a naturally carbonated spring.[8] He was selling "Edanha" bottled water in champagne bottles—beer bottles were too weak to hold its gaseous power—before he sold his first beer from the site. Home Brewing Company of Richmond, Virginia, sold Beaufont Springs Sparkling Carbonated Table Water. This water, from springs in Chesterfield County, Virginia, came in 32-ounce bottles.[9] Many other examples of bottled water sales by breweries are known throughout the history of brewing.

Brande's 1830 discussion of brewing water seems very knowledgeable as to minerals in water and obvious problems of mud, but this was an age that knew nothing of bacteria.[10] "The Thames water, at London, is fattened by the washings of hills and the dirt of sewers, which give it a thick body and muddy taste; and, therefore, it fines well, and makes most drink with less malt." He believed river water to be the best generally for brewing.

Frederick Colyer, a nineteenth century British brewery engineer, architect, and author, stated categorically in 1852 that in the selection of a site for erection of a brewery "the most important [consideration] being the water supply, both as to quantity and quality."[11] In addition to the obvious water that makes up the majority of a beer, water was needed for malting, mashing, yeast washing, keg and bottle washing, and especially for cleaning the brewery proper.

Truth as well as hype were found in the water claims of the past. Insufficient water would preclude a location from being appropriate for a brewery. A large supply of pure water made for an excellent location for a brewery. Ideally, water needed to be hard, but not too hard—water that was colorless, odorless, and without organic substances proved best.[12] Soft water dissolves too much of the phosphates from the barley and was thus generally, but not always, rejected.[13]

In areas where these standards could not be met, brewers used whatever water they had. During the Middle Ages water contaminated with sewage was even used for brewing. Unbeknownst to them, the long boiling probably killed dangerous organisms.

Some brewers apparently did not consider water supply before construction. Henry Mayer, brewery owner of Collinsville, Illinois, reported in his diary in January, 1860, that all three of his wells were dry and he had to haul water from Cantine Creek; by October that year, the creek was also dry and he was hauling water from Clark's pond.[14] His brewery had experienced floods previously. The quality of the water from the creeks and ponds was probably very poor.

Because breweries often had large roofs, brewers could collect rainwater to be stored in cisterns. Colyer said rainwater was not suitable for brewing, but after filtering, was the best for horses.[15]

By early in the twentieth century "Burtonizing," or the artificial production of hard water, to make ale similar to that brewed in Burton, England, was common. Soft water was artificially made for those who brewed Dublin-style stout.[16] Techniques by 1900 for improving water included filtering, aeration, distillation, and treatment with alkaline phosphate, fluorides, limewater, and carbonate of soda. Schlitz, for example, just after 1900, advertised that they went down 1,400 feet for their water.[17]

During a report and discussion of the city water works in 1905, one Boise, Idaho, councilman asked if the local water had ever been analyzed to see if it was pure. The water company had never done that, but the local brewery had. A councilman remarked, "It would look as though the brewery had more regard for the health of the beer drinkers than the city had for the welfare of the people who drink water."[18] A copy of the brewery's water analysis was to be submitted at the next meeting.

Until the 1930s large American breweries were reluctant to brew their old brand at new, distant locations because of differences in the water. Most brewers believed that "chemical processing of water had still not reached the point where water used for brewing from one place to another could be rendered indistinguishable."[19] When Falstaff of St. Louis bought plants in New Orleans and Omaha that decade, and brewed beer indistinguishable from their St. Louis product, the precedent was set for others to follow.[20]

Archaeological Considerations of Water

In an archaeological context, breweries will have a source of water nearby. Generally, a well, quite often a very deep one, should be extant for the location. Physical evidence would be expected. Water pumps, including windmills, might be present at breweries that operate from their own water supply. Storage tanks for water and storage tanks for heated water are also generally found. Some breweries merely used the local tap water.

A brewery could be expected to have water pipes remaining. Hundreds or thousands of feet of pipes are often found in breweries, and some of these are in places where removal would be difficult after the closure of a brewery. For instance, the archaeology done on the Klondike Brewery in Dawson, Canada, found several pieces of pipe measuring 1.25 inches in diameter.[21] The angles of some pipes suggested the water originated in the boiler room and then ran along the south wall. At one point the remaining pipe was bracketed to a wooden post and angled upward through the floor.

Cereals or Grains

Brewers have used barley, rye, oats, wheat, corn, rice, and so on for brewing in some noteworthy amounts.[22] Less commonly, more exotic materials have been used such as buckwheat, wild rice, potatoes, chestnuts, bananas, molasses, sorghum, honey, sugars, maple syrup, millet, and others.[23] An 1830 British book chided, if not condemned, brewers who used licorice root, treacle, or mucilage of flaxseed to replace malt sugar.[24]

The Rise of Barley Farming

The fact that barley makes poor bread may be the reason it was relegated to brewing, while wheat was generally left for bread-making. Barley malt was ideal for brewing, and this fact was recognized over centuries and in many regions.

This ancient grain occurs in numerous varieties—collections with over 50,000 varieties are known—but can generally be differentiated into two categories: two-row and six-row types. The color of the pearl is either blue or white, although the blue may be so weakly developed that it can be confused with white. To identify varieties, these two characteristics are considered, as well as the rough or smooth character of the beard hairs and the length of the ranchilla hairs on each grain.[25] An 1852 book described and differentiated two-row barley, *Hordeum vulgare*, and six-row, *Hordeum hexastichon*, also called "bigg."[26] Bigg thrived better in high, cold climates and ripened more rapidly.

During the nineteenth century, American brewers began to add other grains to the mash tub to make the beer lighter in body, and to compensate for the types of barley found in America. Americans generally grew the six-row type, while Europeans grew the two-row. Chemical differences have been identified, such as a higher nitrogen content and enzymatic activity in six-row.[27] Brewers claimed American barley produced heavier beer than European barley, and the beer needed dilution.

Corn and rice were the two grains most commonly used as adjuncts—nonmalt additions to the beer—for the purpose of making lighter-bodied beers. Corn, traditionally, was ground into grits, but in the late twentieth century, corn processors provided their product in the form of corn syrup.[28]

Six-row barley was much cheaper for brewers to buy. Obviously, farmers could grow many different barley varieties if production was economically feasible and the desired variety thrived in a given region.

When malt became expensive in England due to heavy taxes and low production during the French Revolution, brewers discovered that a greater quantity of wort could be produced from an equal amount of pale malt than from brown malt.[29] They consequently began to use much more pale malt to brew beer and porter. Since the color of the beer was too light, many brewers added burned sugar syrup to color the final product.

Barley was not always a common American crop. Davis Embree, Cincinnati's first brewer, advertised in 1812 that he would pay 62 ½ cents per 47-pound bushel for barley, and he needed 15,000 bushels. A year earlier, he had announced that he had seed available at the brewery and it should be sown at the rate of 1¾ to two bushels to the acre.[30] Beginning in 1815, Joseph Philipson, a St. Louis brewer, advertised free seed to farmers who would grow the barley he needed at his brewery.[31] Examples of this continued in various areas for decades to follow.

In 1877, 1,614,654 acres of barley were cultivated in the United States, producing 34,441,400 bushels worth 70 cents per bushel, for a total value of $24,028,644.[32] That same year, over 10 million bushels of barley had to be imported. The potential for American agriculture was still great.

From Farm to Malt House

Grain was heavy and thousands of bushels of it were needed; transporting the grain to the brewery could be a challenge. For instance, one brewer in the isolated mountain community of Idaho City, Idaho, received 5,000 pounds of barley in early November of 1875.[33] "The grain came in on Emery's fleet of bovine schooners [on the backs of oxen used as pack animals] which traveled between Idaho City and Squaw Creek with no intermediate stops."[34] Brewers in areas with roads, rail, or boat access could more easily obtain barley.

Kessler Brewery of Helena, Montana, received wagonloads of barley malt from Utah, 500 miles away, and many shipments failed to arrive at all during those early years.[35]

When the Overland Brewery in Nampa, Idaho, commenced brewing again in 1934 after the end of Prohibition, it used malt from Manitowoc, Wisconsin, sold by the Rahr Company; local malt was not available. Likewise, when the Pilsner Brewing Company of Idaho Falls began operation in 1935, it bought a train car load of Czechoslovakian malt, since no suitable barley was being grown in the area.[36]

Barley for the malt for the Coors Brewery in Golden, Colorado, was and is grown under contract. In the early 1980s, Coors provided Moravian III seed to Idaho farmers.[37] Upon harvest, the company evaluated the grain to either buy it or reject it. The barley needed to be low in moisture, low in protein, high in plumpness, and light in color. Wet grain ran the risk of spoilage; high-protein barley led to cloudy beer; thin grains fared poorly in the malting room; and dark, tough hulls on the grain resisted sprouting. A general rule of thumb was that one bushel of barley made one 31-gallon barrel of beer. Rejected barley could still be sold on the open market at the going rate instead of the special high rate that Coors paid. This has become the general way the large breweries do business.

Breweries had to store the grain until needed. Large breweries at the beginning of the twentieth century had grain elevators able to hold grain by the trainload. Bucket elevators lifted the grain from the railcar to the elevator. From the elevator, various spiral and belt conveyors moved the grain when ready to begin the malting process.

About 1800, grain was moved through mills and storage houses by a conveyor that had "wooden paddles, the shanks of the paddles being inserted spirally into eight-sided wooden shafts."[38] Later, cast iron flights on wooden shafts were used, then solid shafts. In 1875, H. W. Caldwell of Chicago built an improved spiral conveyor that became widely used[39] (fig. 1.1).

Belt conveyors were used by 1900 to move the grain from storage to malting areas. A belt had moveable "trippers" that enabled the operator to dump the barley where

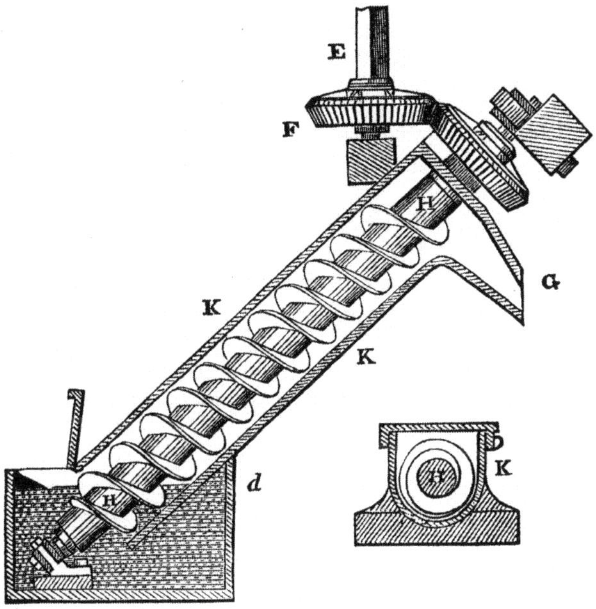

Figure 1.1. Screw to move barley in a brewery. (Byrn [1852] 2002, 58.)

desired. In 1876, the Salem elevator bucket was developed and became nearly universally used.[40] In 1880 W.D. Ewart introduced "link-belting" which he had patented in 1874 and was made by the Link-Belt Machinery Company of Chicago. F. H. C. Mey of Buffalo, New York, patented similar belting in 1878, and the Jeffrey Manufacturing Company of Columbus, Ohio introduced its own version of chain-belting.[41] In Britain, a "Jacob's ladder" was the usual method to raise malt or grist within a brewery.[42] For lateral movement, an Archimedes' screw made of cast iron was still used. The new malt house at the Idaho Falls Brewing Company in 1908 had an elevator and a bucket elevator for hoisting hops, malt, and rice.[43]

Hops

The scales of the female hop blossom have at their base a granular, resinous, bitter substance called lupulin that gives beer a bitter taste, flavor, durability, and a fragrant aroma. Hops (*humulus lupulus*) are in the same plant family as hemp, and grow as a long vine of 20 feet or more.[44] Their use in Europe goes back hundreds of years, and their popularity in England can be traced to the 1400s with great accuracy. Hops vary in aroma, degree of bitterness, and taste components.

Massachusetts was the acknowledged leader in hop production in North America from the middle of the eighteenth century to the early nineteenth century.[45] New York, Wisconsin, and Ohio were once large producers. In the second half of the twentieth century, Washington, Oregon, and Idaho became the exclusive growers of commercial quantities of hops.[46]

A world trade in hops has existed for centuries. American brewers imported large quantities from Germany and Bohemia in the nineteenth century and exported different varieties of hops in large quantities to England and other countries. Breweries in remote western towns often advertised the exclusive use of Bohemian hops. The relatively small quantity of hops needed made importation possible. In 1872, American brewers paid $785,525 for imported hops, and by 1878, $2,152,873.[47] Brewers were using about 30,000,000 pounds per year. American farmers saw great potential.

In April, 1891, Blatz of Milwaukee spent a total of $29,000 on hops; $2,100 on imported Bavarian Saaz hops, $14,800 on domestic hops and $12,100 on hop extracts.[48] American hops were considered inferior or at least lacking the attributes of European varieties such as Hallertauer, Mittelfruh and Saaz.

Hops Farming in the United States

Hop growing was a colorful and volatile business in the nineteenth century. Growers made fortunes one year and faced bankruptcy the next. Prices on the international market fluctuated wildly. Diseases or pests could destroy a crop in days.

Picking hops in the fall required an army of hand laborers. In the West, Native Americans often did the picking. At other times and places, the work force was Chinese.[49] Three issues traditionally dominated hop growers' concerns: market swings and price fluctuations, availability of labor for harvest time, and cleanliness of the picked hops, in the form of an absence of stems and leaves with the hop blossoms.[50]

Fresh hops were 70 to 80 percent water. When dried, 50 pounds of fresh hops became about 15 pounds of kiln-dried hops.[51] Hops were kiln dried for many years and the remains of the kilns have been found in many areas where growing once took place. The archaeology of the hop growing business is as much a part of the brewing story as barley growing. A detailed discussion of the structures and equipment related to hop growing is in the chapter on brewery and related architecture.

In the early nineteenth century, hop growers developed hop extracts. These were lupulin and aromatic oils derived from the hops that were then sealed in cans. Later, hops were processed into pellets or other configurations to make their use easier and to assure a supply in years with poor crops. In the late twentieth century, hops were used in four forms: loose leaf, pellets, extract syrup, and pre-isomerized extract.[52]

International Bitterness Units

Bitterness in beer is measured by International Bitterness Units (IBUs). A higher score for a beer reflects a greater quantity of hops used and more hop acids in the final beer. For instance, American Budweiser has an IBU of 10, Pilsner Urquell from the Czech Republic rates 43, and Guinness Stout is rated 50.[53]

Alternatives to Hops

Different flavoring agents can be substituted for hops, and in the microbrew era, many brewers have experimented with flavorings. Brewers in the nineteenth century West faced precarious supply situations. In Helena, Montana, the Kessler Brewery sometimes used spruce needles or wild grape roots to replace unavailable hops.[54]

A British author of 1830 claimed that disreputable brewers used a long list of items to replace hops, alcohol, natural coloring, and flavoring in their beers.[55] These included opium, henbane, cocculus indicus, grains of paradise, Bohemian rosemary, capsium, cassia buds, ginger, aloes, gentian, quassia, orrice root, horehound, bitter orange peel, and other substances. Likely, some brewers also used such things on occasion in America. Residues of such things might persist in the archaeological record.

Yeast

Despite the mysterious nature of yeast, brewers gained practical wisdom about its use long before its scientific description. Yeast was collected from the fermentation vat when its work was done. During fermentation, the yeast propagated itself and there was always more to take out than had been put in. A brewer observed, "Taken from the fermentation vat, it appears as a thick, pasty mass of light yellow color and having a bitter taste."[56] Brewers often sold or gave away excess yeast for baking purposes or as a health tonic. Brewers yeast is still sold as a dietary supplement.

Monk and brewer Benno Scharl, working near Munich about 1810, noticed his yeast produced a lot of foam, and some settled to the bottom of the vat instead of floating on top.[57] He began to reuse only the yeast that had sunk. With this experiment to inspire them, several other brewers in southern Germany began to isolate yeast that sank to the bottom during fermentation and to reuse it exclusively. Yeast that worked on the top was soon considered ale yeast (*saccharomyces cervisiae*), and yeast that sank to work became lager yeast (*saccharomyces uvarum*).[58] The main categories of beer—ale versus lager—are chiefly differentiated by this yeast distinction to the present time.

Byrn, in his 1852 book, refers to "Bavarian" beer when he describes brewing with bottom-working yeast. This was a common synonym at the time. This tiny Teutonic yeast breed in effect conquered the world. Asian beers, African beers, and Latin American beers imported into United States almost invariably have roots in German yeast. German brewers emigrated around the world, introduced lager beer, and soon dominated the local brewing scene just as they did in the United States. Yeast from Munich made this possible.

Danish chemist Dr. Emil C. Hansen, associated with the Carlsberg brewery near Copenhagen, developed pure yeast by starting with a single cell and cultivating it. In America, William Uihlein of Schlitz, and the Wahl & Henius Company in 1886, introduced pure yeast systems.[59]

Practical Considerations of Yeast

The former owner of a Carson City, Nevada, brewery gave a practical description of yeast-handling in the early twentieth century.[60] All the tubs in the brewery were filled with aging beer in the fall and brewing was stopped for the winter. When brewing resumed in the early spring, a half-barrel (15.5 gallons) of yeast was sent from a brewery in San Francisco. The brewer dumped the yeast into a wooden tub in the cellar, and washed it each morning with water. The water was heavier than the yeast and was drawn off. The yeast was then fed flour made of malted barley. In a short time, the whole tub was filled with a "creamy, frothy fungus growth of yeast cells." Yeast was saved from each brew after that to be reused in the next.

In 1906, a half-barrel of yeast being shipped to the Weisgerber brewery of Lewiston, Idaho, exploded in the stage depot and blew tomato crates across the room.[61] Yeast handling offered practical challenges.

Henry Mayer in December, 1859, wrote in his diary that his brewer went to St. Louis to get yeast from the Pacific Brewery, and Mayer hoped it would be "the last yeast I shall be compelled to get . . ."[62] He no doubt hoped that his brewing would be successful enough to result in an overabundance of yeast that would provide a self-sustaining supply.

Yeast Apparatus

A minimal amount of specialized equipment for the handling of yeast would be found in brewery remains that originated before the middle of the twentieth century.

A device called a yeast strengthener was sold to brewers in the 1890s.[63] A crank on top turned interior blades much as an old-fashioned ice cream maker. Inside was a strainer at the top, and the blades forced the yeast up to the top so the yeast was strained. Perhaps this filtered out impurities or added oxygen, or perhaps it was just a pseudo-scientific idea.

Yeast Beyond the Brewery

In 1941, The Falstaff Brewing Corp. of St. Louis insured its yeast against loss by fire, theft, or similar accident for $1,000,000 with Lloyds of London.[64] It was a publicity stunt but demonstrated the value a brewery put on its yeast culture.

During World War II, the brewing industry turned out 25 million pounds of yeast per year for both military and civilian use. Brewers yeast, the richest natural source of B-complex vitamins, was used to fight pellagra in the southern United States and helped cut cases from 7,000 in 1928 to fewer than 1,000 in 1948.[65] The yeast also contains about 50 percent high-grade protein.

Proportions in a Typical Brew and Recipes for Beers

The exact recipes for brewing beer are as various as for bread. Beer may be thick or thin in body, strong or weak in alcohol, pale to dark in color, highly flavored with hops or other spices, or very mildly flavored if at all. A recipe from 1577 for beer used eight bushels of malt with a half bushel each of wheat meal and ground oats, and 3½ to 4 pounds of hops. A sample recipe from England in 1631 used eight bushels (one "quarter") of malt, three hogsheads of water, 1½ pounds of hops, and yeast.[66] Some of the recipes of that era used nutmeg, sage, honey, ginger, marjoram, mint, cloves, molasses, sugar, nettles, dandelion, spruce essence, or even a raw or a boiled capon. Home brewers of the modern era have used such recipes with limited, if not idiosyncratic, appeal.

Brande's 1830 book described London porter and ale as being made with a quarter (eight bushels) of malt for two barrels (72 gallons, not 62 as with American barrels).[67] Byrn's 1852 book had a section on "Ale Brewing on a Small Scale, Suitable for Brewers in the Interior of the States."[68] For two barrels of ale, or 72 gallons, he recommended nine bushels of malt and 12 pounds of hops, with a possible range of 10 to 14 pounds.

In 1915, a barrel of commercially brewed American beer averaged 35.8 pounds of malt, 12.9 pounds of adjuncts (corn and rice), and 0.65 pounds of hops. In 1940, beer averaged 35.7 pounds of malt, 11.4 pounds of adjuncts, and 0.58 pounds of hops. In 1980, beer averaged 26.7 pounds of malt, 12.3 pounds of adjuncts, and 0.22 pounds of hops.[69] Before microbreweries, beer was getting ever lighter and less bitter—blander, in other words.

Nevada brewery owner Arnold A. Millard recalled the recipe used at his Carson City Brewery. "Sixteen sacks of malt were used, along with five sacks of rice and 20 pounds of hops, to produce 1,600 gallons of beer."[70] The description fails to explain the quantity in the sacks.

The Falls City brewery of Louisville in 1950 described a 600-barrel brew: 16,600 lbs. of six-row malt, 5,000 lbs. of yellow corn grits, 3,000 lbs. of refined grits, and an unspecified amount of domestic and imported hops. Beers over 3.2 percent alcohol also received 25 pounds of soybean flakes as yeast food.[71] Local rival Oertel used 8,600 lbs. of pale malt, 100 lbs. of caramel malt, 3,800 lbs. of refined grits, and 110 pounds of domestic hops in a 280-barrel brew.

Steam Engines, Pumps, and Fuel

The first source of modern mechanical power in breweries was steam power. English breweries were apparently using steam power as early as 1784. The rest of the world lagged behind. Steam engines that developed for many other uses were eventually applied to breweries. Francis Perot used a steam engine in his brewery in Philadelphia in 1820. His first one lasted fifty-three years and was a local curiosity.[72] Steam was used until electrical motors developed to the point that they were competitive. A unique application of science and technology was involved in selecting, installing, using, maintaining, and connecting steam engines to various machines.

In 1880 British brewery architect Frederick Colyer described steam engine usage in breweries.[73] He recommended a more powerful engine than necessary to minimize wear and tear. Beam-type engines were not recommended due to the number of moving parts and large initial cost. Condensing engines offered economy of fuel but used a lot of water. High- and low-pressure engines were recommended only for large breweries due to their lack of simplicity. Horizontal engines were "best for a brewery, all contained on one bed-plate, and with few moving parts."[74] The bed-plate, he explained, to which all parts are attached, must be substantial. Colyer went on to describe engine speed, cylinder design, cross design, connecting rods, crank shaft, main bearing, flywheel, governors, steam pipes, and condenser boxes. The most likely part to remain in an archaeological context would be the foundation under the engine. Colyer recommended a concrete base with brick laid on top. Prevention of vibration was paramount.

An 1856 report from Cincinnati described the steam engine in Geisbauer's brewery.[75] "The engine is simple, compact and cheap. It works as its name indicates, with regularity and without jar or noise. One twelve-horse power, consuming a quarter of a cord of wood, or four bushels of coal in twelve hours, does all the pumping, hoisting, fanning, &c., at Geisbauer's extensive brewery."[76] A few years later, the I. & E. Greenwald Co., engineers, founders, and machinists of Cincinnati, advertised its steam engines for distilleries and breweries with "automatic cut-off and slide valve" construction.[77]

Engine rooms had stone or tile floors, had to be shut off from the rest of the brewery to avoid dust, and, according to Colyer, their walls should be decorated or varnished.[78] Larger breweries usually had more than one steam engine. Coal in America was relatively cheap and was the common fuel where available, but it was dirty and required a large chimney.

One observer of steam power in breweries was impressed but remained sarcastic. "The engine is sixteen horsepower. It can, at the same time, grind the malt, sift it, throw it into the mash tub, let in boiling water that it has made to boil, stir up the malt and water, draw it off, pump it up stairs and throw it into the kettle, heat the kettle of liquid until it boils, throw it out into the coolers, cool it, force and carry it off into vats, ferment it, chafe it, and draw it off beer. With a little practice the engine could be taught to drink the beer."[79]

During the archaeological investigation of the Klondike Brewery, the steam boiler was the most recognizable feature. It had been displaced from its original location but was intact. It was 3 ft. 4 in. in diameter and 7 ft. 4 in. high, and a portion of its former 50-foot smokestack remained on top.[80] The boiler faceplate was removed but inspection marks remained and corresponded to Yukon Government records. The records in 1915 described two "Clyde" type boilers of 45 and 18 horsepower capacities. The large boiler was also described as a "Scotch Marine" and the smaller as a "Vertical Tubular." At this site the archaeologist concluded "of the brewery proper, equipment and features of the steam are most complete. The steam boiler, water storage tank, pumps, and other equipment remain *in situ* or close to their original location."[81]

Breweries required fuel to boil, cool, stir, transport, and perform other tasks. For instance, in Idaho in 1910, the U. S. Census reported the brewers used 4.079 tons of bituminous coal, 1,872 cords of wood, and four barrels of oil.[82] No other fuels were used at that time and place.

Steam engines in breweries were stationary and the equipment they operated was located at various places in the brewery, so the power and motion had to be transmitted. Colyer recommended leather bands instead of ropes for pulleys and wheels.[83] Some of the other equipment involved included collars on shafts, tooth wheel gears, disengaging gears, clutch levers, lubricators attached to each bearing, and plumber blocks.

The M. Schultz Eclipse Engine and Pump Works of Cincinnati advertised its "newest and most substantial small engine now made. The Eclipse Boiler Feeder, acknowledged by engineers to be the "Ne Plus Ultra" of feeders. Double acting fly wheel pumps and direct acting steam pumps, for lifting and forcing water for any purpose, especially designed for use at railway stations, distilleries, breweries, mills, soap and oil works and factories generally. They can also be used as boiler feeders."[84]

Most of these boiler types, however, were destined for obsolescence. The Stevens Point Brewery in Stevens Point, Wisconsin, replaced coal-fired boilers with gas units during the 1960s as part of an upgrade and modernization program.[85]

Electricity and Electric Motors in Breweries

In 1879 the Continental Brewery in Philadelphia was first in the world to be lighted by electricity. Industry everywhere was about to discover the convenience of this new source of power. In 1888, Bullock Electrical Manufacturing Company of Cincinnati installed the first electrical equipment in the George Wiedemann Brewery of Newport, Kentucky.[86] Electricity had replaced steam power, and the brewing industry would never look back. Soon the Park Brewing Company of Brooklyn became the first to adopt electrical power exclusively.[87]

The other dramatic effect of electricity was the introduction of refrigeration, eliminating the need for the harvest of winter ice. The last year in which significant amounts of winter ice were cut and stored was 1925. While the brewing industry napped through Prohibition, electric refrigerators made the ice-cutting business obsolete. The vast majority of brewers had given up natural ice by this time. Electricity had really completed its takeover of the industry.

The Processes Applied to the Raw Materials to Produce the Final Product, Beer

Each step of the brewing process has its own characteristics, pitfalls, and idiosyncrasies. A closer look at these steps will reveal the complexities that drive the brewing process.

The Malting Process

Up to the second half of the nineteenth century, the equipment and techniques for malting were quite simple and had changed little for centuries. In 1852, malting consisted of steeping, couching, flooring, sweating, and kiln drying.[88] Malt houses differed in size, of course, and "with the substitution of steam for hand and horse power . . . the outward appearance of the progressive plants was considerably altered."[89]

Maltsters used simple hand tools including shovels, brooms, and screens. Most malting was done on the ground floor and even below ground level to assure dampness and coolness. The steeping cistern or tank was usually located at one end of the floor and the kiln for drying at the other. Over the steeping tank was the barley loft where barley was stored and could be dropped directly into the tank.

Some malting houses used floors of bare earth. Others later made floors of Portland or Roman cement, or a mixture of stone lime, sand, and ashes. By 1877, for instance, Hamm of St. Paul had cement malting floors.[90] The steeping cisterns were brick, cement, or three-inch oak lined with lead or cement.[91] By 1900 some were made of iron. Some steeping tanks had false bottoms to allow more rapid drainage.

During the nineteenth century, thermometers were introduced into malting to determine the most expedient temperatures to use at each step. In the second half of the century, equipment was developed that saved space and labor, and produced more consistently good malt. This new equipment created pneumatic malting, as opposed to traditional floor malting.

First the grain was sorted to have grains as close as possible to uniform size and weight. By 1900, the grain was washed before steeping, but before this time, it was not. Several washing devices were devised for this. Barley running from the bin into the steeping tank was met with and mixed with a stream of water. The barley absorbed the water, which was replaced with fresh after the first twelve to eighteen hours. By 1900, some malting tanks had a continuous stream of water through the steeping tank, and air was forced through the tank from pipes with holes in the bottom.

Another development at this time was the use of two steeping tanks, one below the other, so that when steeping water was changed, the barley was ventilated and thoroughly stirred. Steeping lasted from twenty-four hours to three days, depending on the preferences of the maltster. A longstanding rule was: "When the corn pressed between thumb and finger continues entire in its husk, it is not sufficiently steeped; when it sheds its flour upon the fingers it is ready. If the substance exudes in the form of a milky juice it shows the grain is spoiled for germination."[92] By 1900, bisulphite of lime, carbolic acid, bleaching powder, or slacked lime was often added to the steeping water to prevent the formation of mold.

After steeping, the barley was put in heaps on the malting floor, sometimes termed the couch-floor, moistened hourly, and stirred with simple hand rakes. Special all-wood malt shovels were sold and used because metal might damage the grains.

In 1889, Adolphus Busch installed in his St. Louis brewery two Prinz (of the Prinz and Rau Company of Milwaukee) Improved Barley Graders and Cockle Separators with the capacity of 500 bushels.[93] Each part of the process was becoming automated.

An American book on brewing from 1852 mentioned that British authors described burying maize (corn) until it sprouted and then digging it up to kiln dry.[94] Dr. Byrn had never heard of such a thing actually being done; corn was simply put in a hogshead with perforations in the bottom and moistened with warm water until it sprouted. This must have been quite uncommon. Most brewers who used corn cooked it a long time before adding it to the mash kettle.

Evolution of Malting

From 1800 to 1900, floor malting systems changed form. "The subterranean malt-cellars with insufficient air and light have mostly disappeared, and overground buildings, airy and well ventilated, have taken their place."[95] Brick or porous loam floors gave way to slate or tile floors, which gave way to cement. A few maltsters used floors of monolith—a mixture of sawdust, mineral carbonates, and chlorides. Another significant development came from the Stead Company of England. It used fans to take hot air from the kilns and blow it through the barley before it entered the steeping tank, and again through the malt before it entered the kiln.

An 1830 description of a malt kiln said a wide, short funnel over the furnace opened under a sheet iron floor on which the sprouted barley was laid. The sheet iron floor was "perforated with a vast number of very small holes, which are so close, as to resemble a sieve."[96] After 1860, kilns were equipped with wire cloth, perforated sheet iron, or steel for their floors.

Nicholas Galland, French maltster and brewer, developed box-malting systems that others adopted. Older malt houses made the transition to the new equipment easily. "This pneumatic system can be easily applied to an ordinary malthouse; all that is required is to place a perforated galvanized iron bottom about eighteen inches above the old germinating floor, the intervening space forming the central flue."[97]

In 1887, Ambrose Plamondon secured the sole right to use all of Jules Saladin's malting patents in the United States. He organized the Saladin Pneumatic Malting Construction Company of Chicago. By 1903, the system as altered by American engineer W. H. Prinz was described in *100 Years of Brewing*. A conveyor carried the cleaned and sorted barley from the storage elevator to the steep tanks, where it remained for forty-eight to seventy-two hours (fig. 1.2). During this time the barley was frequently washed by means of a strong current of water that was forced up from the bottom of the tank. This kept all the steeping barley in motion, and dead and light kernels floated off the top.

From the steep tank the barley was floated through an iron pipe into the germinating compartments, where it was distributed evenly and on level on the perforated floor by means of a rubber hose attached to the pipe. The old steep water drained through the floor into the sewer. The barley was turned while it germinated by an automatic machine that consisted of a frame and numerous turning screws. The bottom of the turning screws was an elastic spring attached to a rubber scraper so that the floor was continuously contacted and every grain was turned. The attemperating apparatus supplied each malting compartment with continuous cool air, while atomizers kept a fine water spray on the barley.

The first of these machines was designed for five-day germination but eventually went to an eight-day process in accord with the old English malting timetable. The Rahr Malting Company of Manitowoc, Wisconsin, was the first factory erected with the eight-day system.[98]

W. H. Prinz of Chicago patented the first automatic and pneumatic malt kiln in March, 1894.[99] The kiln had two or three floors so that drying could range from forty-eight to seventy-two hours. This slow drying system produced mellow malt and a saving in fuel costs. A turning apparatus was built into the kiln, similar to that in the germinating boxes, except it stirred the grain without putting the dry grain from the bottom on top of the wet grain. The kiln traditionally filled with the smoke from the heat source, giving the malt

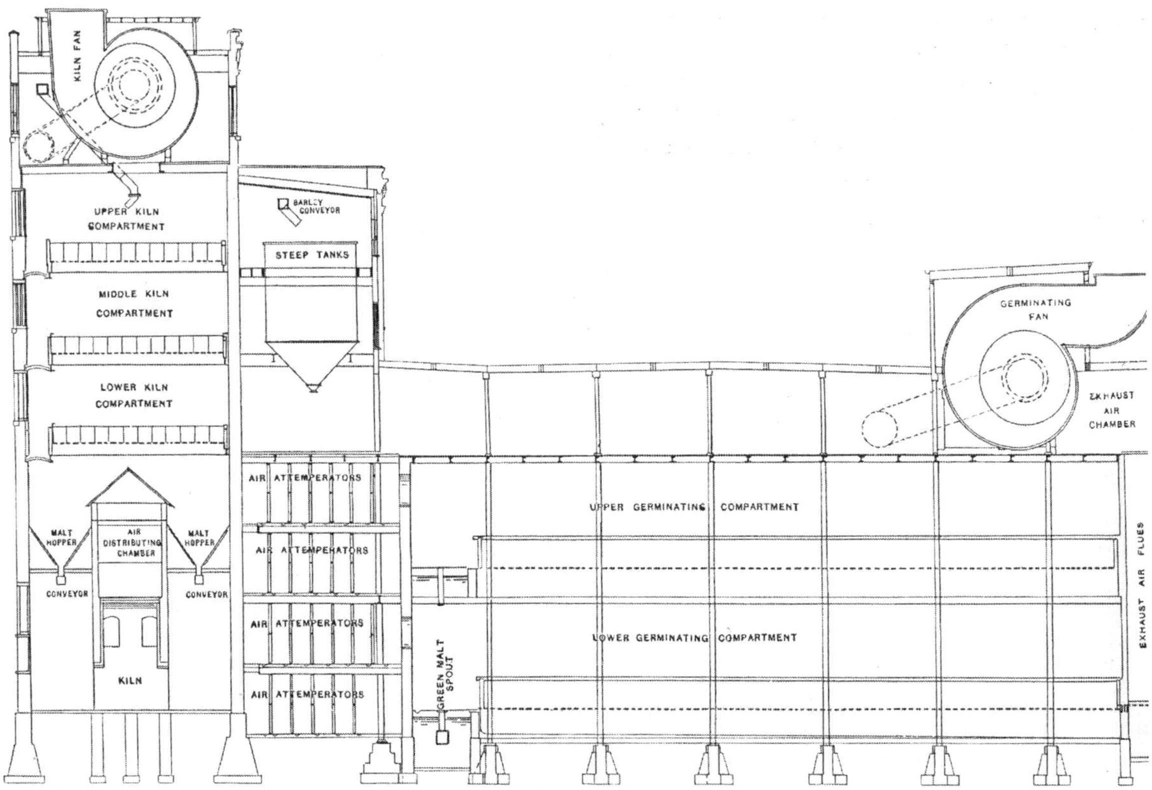

Figure 1.2. Longitudinal section of a malt house constructed on the Saladin-Prinz System. (100 Years of Brewing, 1903, 66.)

a smoky taste. Fuel of beechwood naturally imparted a different flavor than charcoal or coke. Other kiln types began to use the heated air only and allowed the smoke to escape through a chimney without contact with the malt.

In 1881, the first successful dumping kiln floor was introduced in the United States. With this system, the malt could be easily dumped from a higher floor in the kiln to a lower floor and then into a hopper for storage when the kilning was over. Anheuser-Busch installed reversible tray floors in the malt kiln of its new malting plant in 1884.[100] They could dump their entire contents in less than two minutes.

Malt could be very lightly roasted in the kiln to produce pale malt, or darkly roasted to a color approaching black, or stopped at any point in between. Maltsters in America in the late twentieth century measured the degree of the roast on the malt by "Degrees Lovibond," while in Europe, the measure was the "European Brewing Convention" scale.[101]

Other Malting Considerations

In competition with the new pneumatic box malting system were the drum systems. Round drums held one hundred bushels or more of barley and were put on a rack of twelve to fourteen drums. These could be rotated by hand with a crank[102] (fig. 1.3).

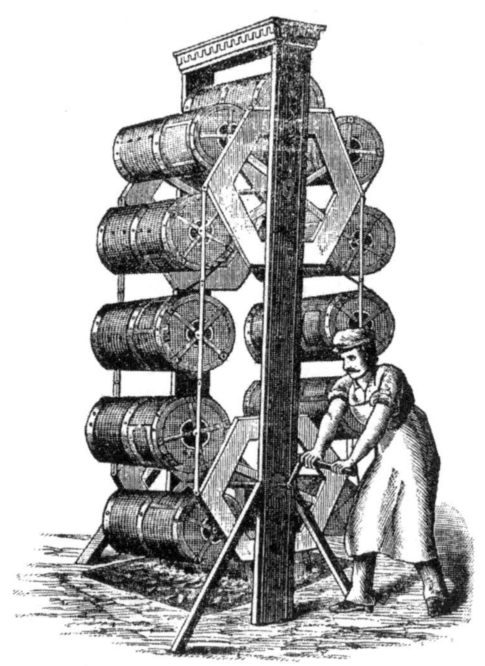

Figure 1.3. Hruby Malting Drum. The laborer turned the crank and all twelve drums in turn went through steeping. (100 Years of Brewing, 1903, 66.)

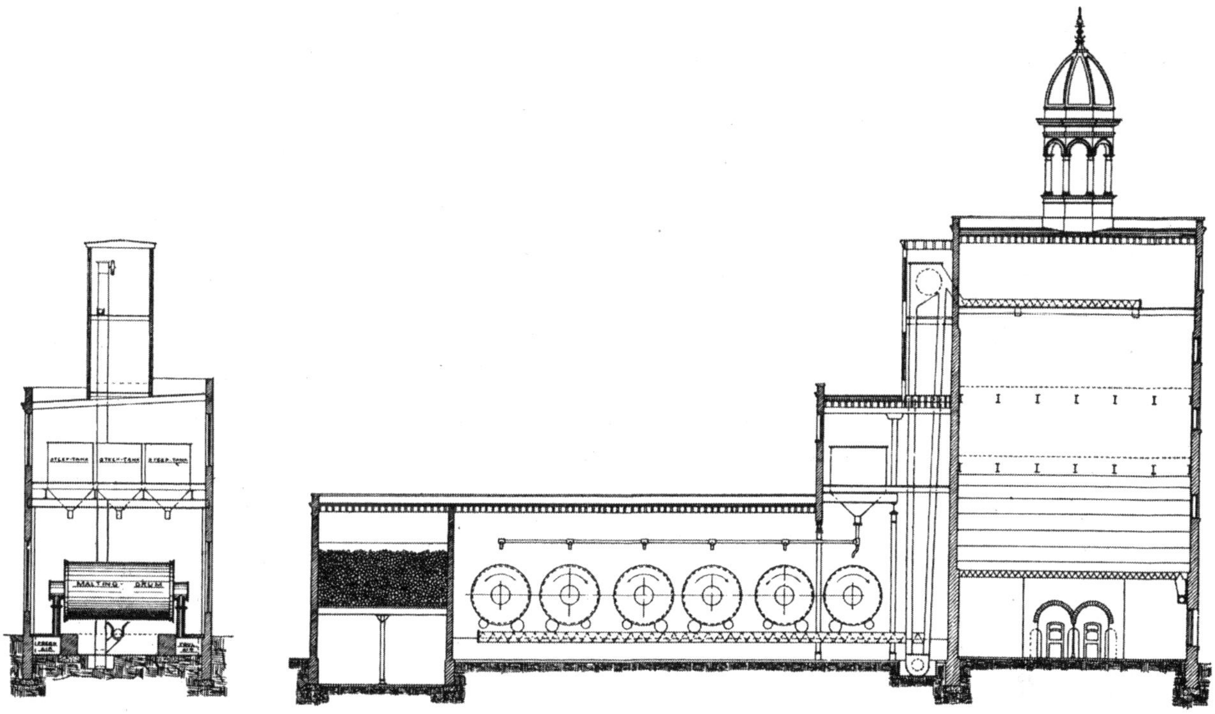

Figure 1.4. The Galland-Henning Six Drum pneumatic malt house. (100 Years of Brewing, 1903, 69.)

The drums had two walls and a number of interior compartments. Once the grain was put into the drum, it stayed until all phases of the malting process were completed. For instance, for steeping, the drums were rotated though water, and for kilning they were placed over the kiln fire and turned. Edward Hruby of Prague, Bohemia, patented this system in 1879. The first such plant in the United States was in 1889 at the Kraus-Merkel malt house in Milwaukee.[103]

Many other malting systems competed, such as the Galland-Henning pneumatic kiln system, the Tilden Drum-Malting System, the Stier Drum-Malting System, and the Dornfeld Malting System (fig. 1.4).

The final step was cleaning the malt in order to remove the rootlets so the malt again looked like unsprouted grain. In old times, the maltster walked on the malt with wooden shoes. In another technique, the malt was rolled down a screen while the maltster rubbed the grain with a cloth-covered hand. Later, machinery performed this task. Machinery was also developed to collect the "dust," the parts removed from the grain.

Some maltsters used special maltster's shoes. "The work is first class, solid, oval, compact and on the sole of the Shoe is fastened the finest rubber (Horseshoe form), connected with nickel plated strips, thus making them very durable. If desired the Shoe can be so constructed as to use the string for closing them up tightly to the ankle, thus preventing the malt scums from getting into the Shoe."[104] Large-scale brewers believed the substantial investment in malting equipment was well spent and eventually made malt production more economical.[105]

Another concern of the brewer was pests. Malt that was stored for some time could be susceptible to weevils. British author W. Brande in 1830 believed that making light malts was the source of the problem because not enough heat to kill the pest was applied. He recommended screening the malt repeatedly to physically remove the insects.[106]

Some breweries bought malt rather than making it themselves, and the separate manufacture of malt at other facilities has been common. According to a document of 1661, the people of New Amsterdam purchased malt from New England.[107] The Cleveland Brewery in 1833 advertised malt for sale to distillers; it also offered to malt the distillers' own grain "on reasonable terms."[108] In 1907, the newly formed malting company in Idaho Falls had $50,000 in capital and was located adjacent to the new brewery. They intended to let contracts to farmers for barley for the next growing season. "The institution should be very profitable as all brewers in the west at this time secure their malt from eastern parts, and the territory to be covered by the Idaho Falls plant will in consequence be very large."[109] In the early twenty-first century, Anheuser-Busch, for instance, operated a giant malting facility in Idaho Falls, Idaho, although the nearest brewery it operated was hundreds of miles away.

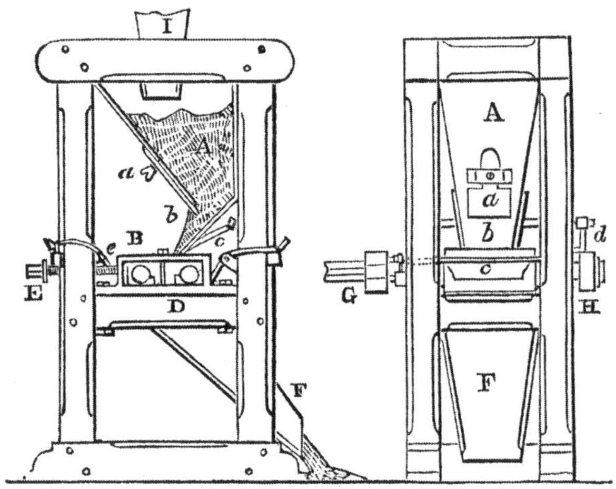

Figure 1.5. A malt mill. Malt was ground before mashing; such flour can be explosive. (Byrn [1852] 2002, 23.)

The Mashing Process

Malt had to be ground into grist, or rough flour when mashing was to begin (fig. 1.5). Mill stones were used for this traditionally, but specialized malt mills replaced them. An 1816 report on the Embree brewery of Cincinnati described a treadle mill "turned by horses and grinds one hundred and twenty bushels of malt a day."[110] Brewer Henry Mayer, writing in his diary in December, 1857, said, "Yesterday whilst in St. Louis contracted for a cast iron mill to grind the Barley to make malt & beer to cost $125 . . ."[111] Nickolas Kessler of Helena, Montana, had a malt mill made by the local foundry and machine shop when first outfitting his brewery in 1864. It was driven by a one-horse sweep; one horse walking around and around, turning the brass gears that caused the small rolls to work in unison.[112] In 1880, Frederick Colyer described malt mills that had changed little, if any, for years and were made of equal-sized rollers of steel or cast iron.[113] A screen was placed in front of these to keep out foreign objects such as stones.

The Koeninger Brothers' Brewery in Ketchum, Idaho, in 1884 put a turbine wheel on Trail Creek to run the barley cracker (mill) and pump water into the vats.[114] Old sources of power were still being used in outlying areas (fig. 1.6).

In 1891, Kaestner's nonexplosive malt mill was advertised through a brewer's supply catalog.[115] The rollers were fitted with a dust-proof steel case and an elaborate system to receive the ground malt. "From this it is readily understood that any explosions taking place must be confined to that chamber of the revolving hopper which happens to be below the grinding rollers at the time."[116] Most brewery fires were attributed to boiler problems, but this was obviously another source, and brewery people were trying to make a safe machine.

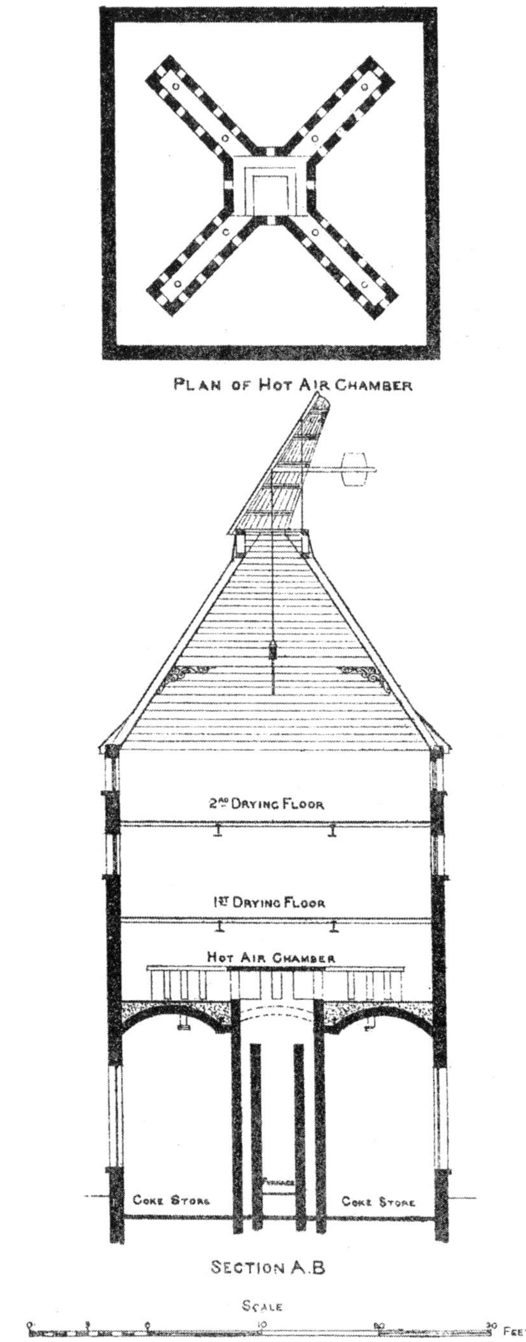

Figure1.6. German malting system as of 1880. (Colyer, 1880, plate 20.)

Figure 1.7. A worker stirring the mash kettle with a long oar. (100 Years of Brewing, 1903, 77A.)

Traditionally, ground malt was added to a large wooden tub, hot water was poured on it, and a laborer stirred the tub with a long, oar-shaped paddle. As many as 160 bushels of malt in a mash tub always employed hand labor, while larger operations might have mechanical stirrers.[117] The tub had a hole in the middle of the bottom that was plugged with a round shaft of wood, called a tap tree, until ready to be emptied. In Britain in 1880, mash tuns (250-gallon barrels) for beer were generally of oak with Dantzic timber bottoms, and those at porter and stout breweries were iron.[118]

Brewery engineer Frederick Colyer also mentions cast iron tuns lined with copper that were heated by steam between the two metals. One of the first mashing machines was invented by W. M. Ellis of New Jersey, and installed in a Philadelphia brewery about 1800. In Europe, jacketed mash tubs so they could be warmed with steam were invented, but in America these did not catch on. In 1828, a new brewery built by John Beveridge & Company at Newburg, New York, had steam-powered rakes in the mashing tub.[119] In 1878, Charles Kaestner of Chicago took out his first patent on an improved mashing machine[120] (fig. 1.7).

Another improvement was making mash tubs with false bottoms. Water could be put in the tank from the bottom by use of the so-called pfaff. Devices called spargers were used to squirt hot water on the grain to rinse out the last bit of fermentable material. "The amount of fermentable sugars and therefore the alcohol content can be varied according to the temperature to which the mash is raised and how long it is held there."[121] Modern breweries often pump the mash tub contents into a lauter tun, also called an underback, where the sweet liquid is separated from the grain residue. Historically, these two processes were often combined.

In January, 1860, Henry Mayer, owner of a small Illinois brewery, recorded in his diary, "I now have the Mash Tub enlarged & a new iron bottom; it suits well, now the juice of the malt is as bright and clear as Wine. . ."[122]

The description of mashing from the Carson City, Nevada, brewery gave a practical perspective:

> The rice was boiled in a metal tank for some length of time. The malt was ground up in a malt mill and emptied into the mash tub where it was cooked by steam. It was necessary to stir this at all phases of operations. Within the mash tub was a revolving device which was affixed with paddles and agitators. At the proper time a valve was opened and the cooked rice was dumped into the mash tub. Both grains were further cooked. The power was turned off and the grains were allowed to settle. The liquid was then drawn off into the copper brew kettle. The grains was sprayed with hot water from an interior pipe ring to assure complete recovery of all the nutriment from the spent grains.[123]

The question arises as to how the mash kettle was kept hot before the days when steam jackets were put around it. In many areas a decoction system was used in which some of the mash liquid was drawn into the brew kettle, boiled, and then put back into the mash. This was repeated to keep the temperature up. This was called decoction and took a long time. *One Hundred Years of Brewing* blames this system for causing nineteenth century brewery workers to often work 18-hour days.[124]

Later, the infusion method involved blending the water and malt grist at low temperature, then heating it and having a final quick boil. This took much less time and a brew begun at 3 A.M. could be in the cellar fermenting by the afternoon. Hence, workers could be assigned "rational working hours."[125] The liquid that left the mash tub at that point was wort, and ready for brewing (boiling).

Boiling, Hopping, and Cooling the Wort

In old times, the pan or kettle in which the wort was boiled was the only copper vessel in the brewery, and became known as the "copper." In small early breweries, one copper heated water, boiled the mash during the mashing process, and then boiled the wort. Larger breweries began to use two kettles, the first known as the hot-water tub, and the second as the brew kettle. Brew kettles could be round, as they were in England, or square, as in Germany (fig. 1.8). They could be covered or open, and they could be curved inward at the bottom if they were to be emptied by gravity, or curved outward if they were to be manually dipped out. If the kettle

Figure 1.8. J. Adam Lemp's original brew kettle used in St. Louis in the 1840s. It is still owned by the Missouri Historical Society. (100 Years of Brewing, 1903, 212.)

was heated by having a direct fire under it, the bottom had to be especially strong. This was the only system before the introduction of steam brewing.

The brew kettle used by Adam Lemp when he opened his first brewery in St. Louis in 1840 is extant as of this writing and a fine description of it is available.[126] The kettle was probably constructed in Chicago or St. Louis. It held 372 gallons, the equivalent of twelve barrels. The kettle was made of eight copper sheets hand beaten into place; in this era before copper could be welded, it was also riveted. The seams were lapped and the bottom was one single sheet. Two spouts were plumbed into the kettle, which stands about 4½ feet high with an 18-foot circumference, and weighs between 550 and 600 pounds.[127] It sat on a foundation of brick with an open center where a fire to heat the kettle could be stoked. By way of contrast, in 1892, the Lemp brewery had two 500-barrel, steam heated kettles.

Richard G. Owens, a Wisconsin brewer of Welsh descent, began brewing with a wooden box lined with copper sheeting; later, he drove a horse team to Chicago to procure a 12-barrel kettle. In 1845, a coppersmith arrived in Milwaukee and made a 40-barrel kettle for him.[128] How does one boil liquid in a copper-lined wooden box? Possibly by adding red-hot stones, or running tubing filled with steam or heated water through the kettle. A European tradition of "flaming stone" beers known as *rauchenfels Steinbier* once existed, made by putting hot stones in the kettle.[129] The malt sugar caramelizes to the stones and creates a unique flavor.

Steam power was already in use in breweries and was often used on mash kettles before it occurred to anyone to boil the wort with steam. Brewers first lowered steam coils into the wort to bring it to a boiling temperature. Later, steel jackets with flues were made that surrounded the kettles. After this development, virtually all kettles were round and covered,[130] and most of this transition was dated from 1870 to 1900.[131] In 1880, Colyer said steam coil heating was passé and the costs of the various other systems nearly equal.[132] A brewer's supply catalog of 1891 from New York showed a closed-top steam kettle that was round and sat on a metal stand.[133]

Two examples stand out of the significance of the brew kettle. Fred Mayer of Collinsville, Illinois, sold his brewery property in May, 1869, for $1,900. In addition to the real estate, the mortgage specified "one large copper kettle and one mash tub on said premises."[134] Louis Abegg, a nineteenth century Swiss émigré who worked briefly as a brewer in Collinsville, came from a brewing family and brought his small copper kettle to America with him. His descendants still own it but use it to make apple butter, not beer.[135]

The purpose of boiling the wort was to reduce it to its proper strength and extract "flavor, taste and keeping quality" from the hops.[136] Two to four pounds of hops were added per barrel for ales, and not less than four pounds for porter. Each brewer had his own formula for how long to boil and how much to add.

The account from the Carson City, Nevada, brewery explained the technique. "This liquid [from the mash tub] was then boiled in the brew kettle for approximately five hours. About 20 minutes before this was drawn off into a large metal tank to cool, hops were added."[137]

After boiling, the hops had to be filtered out of the wort. Brewers once did this by running the hot wort through a basket lined with straw. A filter device called a hop jack (or hop back or jack back) was developed to filter the hot wort, and became universally used.

Yeast could not be added to wort near the boiling point. The wort had to be cooled as quickly as possible to prevent infection and the formation of undesirable acids. For generations, brewers ran the wort into shallow vats for cooling. One of the first breweries in St. Louis cooled the wort in a dug-out canoe—called a piroque—on the outside of the north wall.[138] Most brewers used less primitive techniques.

Pipes were later introduced into the shallow vats and cool water was run through the pipes to hasten cooling. An inverse system was to run the wort in pipes that went through vats of cold water to transfer the heat. Another technique was to spray the hot wort into the tubs to air-cool it. All these techniques were abandoned when the Baudelot "refrigerator" with its ammonia-based system was introduced. Numerous companies made similar wort coolers.

The Fermentation and Aging Processes

Brewers controlled fermentation and yeast activity with the thermometer and the saccharometer.[139] These replaced touch and taste as ways to determine the timing of fermentation. Principal fermentation converted saccharine matter to alcohol (51.1%) and carbonic acid (48.9%), and produced the regeneration of yeast. Secondary fermentation happened in the storage vats slowly, and ripened the beer and eliminated impurities. The last stage was fining, in which the beer acquired full flavor and carbonation, and was ready to ship.

One of the few physical objects brewers used for fermentation before artificial refrigeration was the attemperator. These were boxes that were filled with ice and floated in the wort to cool it and to secure a uniform fermentation throughout the tub. Fermentation generates heat, and this helped to modulate it.

To prevent contamination, Casper Pfaudler developed a system of fermentation of wort in a partial vacuum.[140] Fermentation was done in a glass-enameled tank in which sterilized air was admitted from time to time and carbonic acid was drawn off the top. Fermentation by this method was faster, saved labor and space, and avoided contamination. Almost all breweries now ferment in closed vessels.[141] Fermentation time varies greatly. Ales ferment in five to ten days, whereas lagers require fifteen to twenty-two days.[142] Modern chemistry can speed these processes.

Byrn in 1852 termed the next phase of brewing "cleansing."[143] If the yeast was allowed to stay in the fermenting tun after it stopped working, it would gradually mix with the newly fermented ale and cause a disagreeable bitter taste called "yeast bitter." To prevent this, the ale was put into casks. The bung was not yet put in the casks so that any newly formed yeast would rise to the top and flow out. This prevented the undesirable continuation of the fermentation.

The beer was then aged in huge vats. Ales might be allowed to mature for a few days; lagers can be left to mature in a cold environment from two weeks to several months (fig. 1.9). Schlitz, about 1900, advertised that its beer would not create biliousness because it was aged for months in glass enameled steel tanks.[144] Walter's Brewery in Eau Claire, Wisconsin, used cypress aging tanks until 1970, when they were sold to a Canadian food processor and replaced with stainless steel tanks.[145]

The third step in fermentation is the fining. The beer is drawn out of the aging casks into the chip casks. Thin chips of beech or maple wood are put into these large vats and the yeast particles and other impurities cling to the wood, creating a clear, sparkling final product. The alternate method to clarify the beer is to add isinglass. This is a gelatinous substance made from the air bladders of sturgeon. A small amount is spread throughout the beer and attracts sediment.

Next came krausening and bunging, the final steps before the beer was ready for market. In krausening, a small amount of fresh wort was put in the final keg to start a secondary fermentation and cause carbonation. This is much the way that champagne is still carbonated. Zoller & Schimper patented a Kraeuse apparatus in 1887 that was basically a gauge to measure the exact "quantity of Kraeusen Beer required for the shaving casks, for mixing quantities of different brewings in color, age and strength."[146] Modern brewers have essentially abandoned this process and impregnate the brew with carbon dioxide under pressure to carbonate the beer.

Figure 1.9. Brewery aging vat in the ghost town of Silver City, Idaho. It was put in the middle of the street, roofed, and kept as a local curiosity for decades. (Photo by author.)

Brewers used bungs with vents to prevent explosions caused by excessive pressure. Vented beer bungs, high and half bungs, and double flange bung hole bushings were all sold to brewers. Bungs are frequently seen all around microbreweries and could easily be part of the archaeological record. Several barrels full of bungs were found at the Gilbert Brewery in Virginia City, Montana.

Steam beer was an American invention that originated in California but spread throughout the western mining area.[147] Lager was brewed in a traditional way except the brewer had no way to cool it properly. Fermentation took place at a higher than ideal temperature, resulting in a high level of carbonation, and the beer built enormous pressure in the barrels that were delivered to the saloons. If not carefully opened, the beer would shoot out as "steam." The description of the exact brewing technique used for steam beer from Carson City, Nevada, is instructive. "When properly brewed [boiled] it [the wort] was drawn off into a 'ship.' The ship was a shallow metal tray sufficiently large to hold the brew. Metal tubs called 'floaters' were filled with ice to cool down the beer. When the beer had settled and cooled to the proper temperature, it was drawn off into the beer barrels. The steam beer was naturally carbonated by the addition of a certain amount of brewers yeast in each barrel. . . . It was held for a certain length of time to ferment and build up pressure in the barrel. Deliveries were then made."[148]

Racking and Bottling

Racking is the process of putting the beer into shipping casks or bottles. American beer consumers in the nineteenth century and since demanded more than a pleasing taste; they wanted a pale color with luster and clearness, effervescence, and of course, they wanted it served cold. To achieve these complex objectives, brewers developed many innovations to please their customers.

One of the simpler steps in this process was the development of filters to be used as the beer was put into the final kegs before leaving the brewery. A device called a beer observer, patented in the early 1890s, was attached to the racking faucet, and with the aid of a candle, the brewer could see the beer flow into the keg "without continually sampling, to observe whether the beer runs clear and brilliant from the cask."[149]

Beginning in the late 1870s, bottled beer was generally pasteurized before leaving the brewery. By 1900 some of the keg beer was also pasteurized. Some American brewers, most notably Coors, have never pasteurized their beer. They have claimed their filtering system achieved the same end.

Racking troughs were used to speed the filling process. The device sat on wheels for mobility and a large hose from a vast storage tank was attached. Four kegs could be filled simultaneously from the four faucets fed by the hose. These were made of wood with iron corners, or all iron.[150]

Barrels of beer that had aged in a cellar had to be raised up to ground level for shipping. Brewery engineer Colyer in his 1880 book devoted an entire chapter to hoisting equipment. The endless chain lift was the simplest device for raising beer from the cellar.[151] Many other competitors existed.

Bottling on a very small scale was probably practiced in America during the seventeenth century. By the mid-1850s, before the lager revolution was relevant, Cincinnati had at least three nonbrewery bottlers of ale and porter. By 1875 one independent bottler there, Star Lager Beer Bottling Company, had a 40-foot by 100-foot, five-story building and employed forty to fifty workers.[152] Brewers also set up auxiliary bottling companies or put in bottling departments at the brewery. By 1895, twelve of the sixteen beer bottlers in Cincinnati were part of a given brewery.[153]

Until 1890 antiquated tax laws required beer to be put into a keg, tax stamps affixed and cancelled, and transported over a public road—although a mere city alley qualified—before it could be bottled. Captain Pabst of the Milwaukee brewery lobbied and had this law amended so that a pipeline could send the beer directly to the bottling house. The pipes had gauges that a government agent could lock and seal so that no untaxed beer could be bottled. These pipes were also constructed so they could be cleaned with steam. Brewers had to submit a diagram showing all pipes, cisterns, vats, tanks, valves, cocks, faucets, and gauges. These documents could be valuable to an archaeologist.

All breweries doing large amounts of bottling soon switched to the new method. In 1911, 15 percent of breweries had pipelines to their bottling houses; by 1917, 29 percent had them—a definite trend.[157]

In 1907, Kessler Brewery of Helena, Montana, had a government-issued meter to measure beer production for tax purposes installed in the facility, the first in the region, or so they claimed. They wanted to increase their bottled beer production and distribute the beer over a wider marketing area.[158]

Large shipping breweries led the way in pushing bottled beer. Martin Stack identified five reasons for this. First, it allowed breweries to decrease reliance on saloons. Next, it could be shipped to dry counties and states, and it facilitated efforts to create brand loyalty. Bottled beer could also be targeted to affluent, middle-class consumers. Finally, it offered higher profit rates for the breweries.[159]

Beer Consumption by Brewery Workers

A question that always arose regarding the taxation of beer at the brewery involved the beer consumed on the premises; this was not subject to tax. Tax collectors were incredulous about the large amounts that were claimed to be tax-free. The Wahl-Henius institute in 1902 found that brewery workers averaged drinking 25.73 glasses of fringe benefit beer daily per capita.[154] In 1887 a Lander, Wyoming, brewer filled out the classic tax report to the government: "During the month of September I brewed three barrels of beer, I sold one and drank two."[155] Lou Bosler, who worked in shipping at Fehr's Brewery in Louisville from 1951 to 1963 said about half the workers drank a lot of free beer and half did not drink any. One small man in the bottle shop drank forty-two bottles in an eight-hour shift.[156]

Disposal of Spent Grain, Spent Hops, and Excess Yeast

After brewing was finished, mashed grain, used hops, and excess yeast remained. Brewers had to dispose of these efficiently, and they developed equipment to aid in this. After the hops were filtered out of the wort by the hop jack, they were put into a hop press to squeeze out the liquid and convert the hops into a solid mass. Early types used a simple screw to squeeze the hops; later types used hydraulic power.[160]

Yeast presses worked much the same way. The yeast was pumped into bags that were laid between the plates of the press. The screw squeezed out the liquid, resulting in dry cakes of yeast.[161]

Spent grain was sometimes sold wet, or used on premises at those breweries that fed animals. Later, spent grain was dried and bagged for sale as animal feed.

Brewery Vessels

In medieval times a brewer usually was also a cooper and made his own casks, tuns, tanks, and tubs. By the nineteenth century, brewers generally bought their vessels from cooper shops and did only repair work at the brewery. The owner of a small brewery in 1858 recorded: ". . . he [Mr. Beck] helped me move the Hen House in the Stable yard and fitted it up as a Cooper Shop."[162] An 1891 brewer's supply catalog offered storage casks, fermenting tubs, and all sizes of barrels, all proudly made of split oak exclusively.[163] By 1900, giant breweries found it economical and convenient to have their own large cooperages.

Anatomy of a Barrel

Oak was the most common wood used for casks and barrels, but some vessels in America employed cedar, cypress, or California redwood. Vessels were coated inside with pitch that by the 1890s was slowly giving way to varnish. This was to prevent the beer from contacting the wood and acquiring its taste.

Originally, the cooper worked by setting the pitch on fire inside the barrel and then rolling and turning the barrel to coat the inside evenly. Pitch was a resin from various firs and pines. The resin was boiled until all turpentine smell was gone. While still warm, it was cast into blocks and packed in barrels or boxes. Good pitch imparted virtually no flavor to the beer.

In 1864 Matheus Gottfried of Chicago patented a device to use hot air instead of fire in the pitching. In1891 the Stromberg Fire Pitching Machine was still sold.[164] These came with either hand-powered blowers or steam-powered blowers.

Henry Mayer, a Collinsville, Illinois, brewery owner, described the work of preparing casks and barrels in his diary. "23 January 1860. Nearly all casks are in the big cellar; they have been a prodigious long time at it; John [his brewer] told me at first, he could rosin 20 to 23 Big Barrels a day, instead of which 9 was about the average; I have no use for the 4 Big Hogsheads that Mueller has to bring, although John says he will have need of them in the cellar to use as tubs."[165]

A Madison, Wisconsin, cooper, Frank J. Hess, explained his manufacturing system as it was about 1900.[166] He employed thirty-five men and got his white oak staves from a sawmill in Arkansas, but his ends were Wisconsin oak. The staves had to be cured outdoors for a year and a half, or they would crack when bent. Half barrels weighed as much as 80 pounds when finished. Hess explained the importance of cattails, the residue of which might be found in the excavation of a cooperage:

> We put cattail leaf between every stave. This helps to keep the barrel from leaking. A tavern empties a barrel and rolls it outside where it sits in the sun. It dries out, and it could leak when refilled. When breweries pick up the barrel they sprinkle water on it and it expands. All the while the cattail leaves between the staves make sure the barrel doesn't leak. Other materials might rot, but not cattail. We used only the leaves from the female cattails, not those with the cigars. We got our cattails right here in Madison, from where the Dane County Coliseum is now and from where the new city bus depot is. Those were marshy places where cattails grew. We harvested them in the fall and hung them up. It took a year for them to dry.[167]

The Cooperage Industry

By 1859, Cincinnati could boast 130 cooperage shops employing 1,756 workers and creating revenue of $1,510,000. In 1889 Louis H. Buhrman of that city advertised that he could supply 1,500 beer packages daily.[168] Louisville had twenty-three cooperages in 1880 and twenty in 1905. By 1891 steam power assisted the labors in most of these.[169]

Coopers made barrels for an enormous number of commodities before modern packaging developed. Barrels made for dry goods might be made of pine or other soft wood and did not show precise workmanship. These were called "slack" barrels.

Barrels made to hold liquids were called "tight" barrels; these were generally made of white oak or beech that had been aged a year and kiln dried for two weeks before being cut into staves. The staves had to be carefully and uniformly tapered to fit tightly at the ends and have a proper bulge in the middle. Assembly of the staves was called "raising the barrel." The staves were set vertically in a ring with a temporary hoop in the bottom, then they were steamed, drawn into the final shape, fastened with temporary hoops, and finally redried. A groove called a croze was cut near each end for the head pieces. The permanent end hoops were put into place next, then the bilge hoops near the middle, and the quarter hoops between the bilge and end hoops. The bung hole was drilled in the side to complete the new barrel.[171] In some locales, the barrels made for wet goods had alder hoops, and barrels that had to stand on end had thicker, sturdier chimes (the part of the stave extending beyond the head).[172]

Often, brewers in the early days in the West used oddly sized barrels of 5 or 10 gallons. For instance, a deed in the records of the Weisgerber Brewery showed that Ernst Weisgerber bought the California Brewery in Lewiston, Idaho, from Charles Brown on October 16, 1863, for $2,600, with Brown retaining the rights to twenty-four ten-gallon kegs.[173] Brewer Brown had been selling beer in odd-sized kegs.

Maintenance of Barrels

Wooden barrels had to be worked over by a cooper when they were returned from dispensing beer. They had to be

Beer Barrel Sizes

The standard American beer barrel was and is 31 gallons. These are so large and heavy that they are rarely used. Half barrels (15.5 gallons) were commonly used for saloons, while quarter barrels (7.75 gallons) were used for private parties and small establishments.

In the eastern United States, one-sixth and one-third barrels were used. Barrels from different locations often differed slightly in their outside dimensions. A quarter barrel from St. Louis, Rochester, and Denver was 12.75 inches across the head, 19.5 inches high, and 51 inches in circumference in the belly. New York and Chicago quarter barrels were 12.25 inches across the head, 20 inches high, and 51 inches in circumference at the widest girth.[170] These differences might aid an archaeologist in identification.

inspected, rinsed, and then pitched, which sealed and sterilized the interior. A brewer from Nevada explained:

> The pitching kettle was a half round kettle under which a fire was placed. Pitch came in blocks, was broken up, put in the kettle and heated to 350 degrees Fahrenheit. On the enclosed top of the kettle there was a saddle which held the barrel with the bunghole on the side down over a spray type nozzle. A hand operated pump in the bottom of the kettle sprayed the inside of the barrel with a coating of pitch. The barrel was allowed to drain and then taken off and rolled on a framework into the washhouse where it was stacked and sprinkled with water until tight. It was again rinsed, inspected, and when a cork was hammered in the end of the barrel it was ready to be filled. The name of the brewery had to be burned or stamped on the top of the barrel, and it was illegal to fill the barrel of another brewery.[174]

Small breweries used a brand much like a cattle brand to burn their names on the kegs. Big breweries had a complex machine for this. William Bartholomay's Patent keg branding machine was sold in 1891.[175] It had two heaters, a compressed air tank, adjustments to accommodate any size of barrel, easily removable branding plates, and was said to last for years (fig. 1.10).

A trough-style keg sprinkler was developed to keep kegs moist, in which kegs sat in a small metal stand in a wheeled trough, and a sprinkler underneath sprayed them.[176] The sprayed water ran back into the trough and was reused. The sales pitch for this equipment was that this saved water for brewers who had a short supply, and it could be completely drained in winter to prevent freezing. Another style had two pipes leading to the sprinkler stand so that hot or cold water could be selected for sprinkling.[177]

Figure 1.10. Koeninger Brewery, Ketchum, Idaho (Sun Valley) was branded on the end of this keg. (Photo by author.)

Breweries also used automatic keg scrubbers. For instance, during the 1890s Pohl's patent keg scrubber had brushes for each end of the keg and a long brush to clean the side.[178] A keg was put inside the heavy metal frame and rollers moved it around so the brushes worked on the entire outer surface.

A cooper had endless work about a brewery. Future brewery owner John Hendel first appeared on the Idaho scene in 1886 when he was working in Boise at Brodbeck's brewery as a cooper. At one time Hendel built a tub on the second floor that held seventy-three barrels of water. It was made of 2-inch planks, was filled at night, featured automatic shutoff when full, and furnished water to the brewery all day.[179]

The book *100 Years of Brewing* reveals the danger in varnishing casks and the need for safety in ventilation and avoiding open flames. An example from my research confirms this. Franz Koehler, who later became a brewery owner in his own right, worked as a foreman at Brodbeck's Boise, Idaho, brewery in 1890. He was varnishing the inside of a huge vat when someone handed him a candle so he could inspect his work. The fumes caught fire and Koehler was in an inferno. A bystander threw in a rope and dragged him out. Amazingly, he was not seriously injured.[180] *The Boise Statesman* article went on to lecture the brewery trade about using good varnish with low alcohol content and keeping candles away. Sound advice, but not timely.

A device called an air injector was invented so that a supply of air was pumped into the cask while it was being varnished to prevent explosions. The Funck brand advertised with a picture of a worker standing inside a cask and varnishing by the light of an electric bulb.[181]

Barrels could fail and cost a brewer a great deal. Henry Mayer of Collinsville, Illinois, wrote in his diary on July 12, 1859: "I discovered in the Cave today that the end hoops of one large 20 brl. Cask came off and the entire contents leaked out, that is a loss of over 160$."[182]

In 1907, the Bannock Brewery of Idaho Falls, Idaho, offered surplus equipment for sale including thirteen oak tubs from 8 to 15 barrel capacity, and twenty-three oak casks from 10 to 15 barrel capacity.[183] The basic equipment of wood vessels persisted through the whole nineteenth century.

In 1903, the first glass-lined storage tanks were installed in the Kessler Brewery in Helena, Montana, to replace the old wooden ones, which were too small to keep up with increased demand.[184]

Wooden barrels did not quickly become obsolete. In the 1937 *Brewery Age Buyer's Guide* forty-eight manufacturers of wooden barrels and kegs were still listed.[185] The Hess cooperage of Wisconsin said good labor was easy to find during the Depression when legal beer returned but still he could not meet the demand, and that is what forced brewers to use metal kegs.[186]

Once a keg of beer reached a retailer, special, identifiable equipment was still needed. Some saloons had cellars where the barrels or kegs were set both before and during use. Many had no such facility and used specially designed iceboxes. For instance, a 90-by-65-by-32-inch cooler weighing 32,700 pounds held two kegs to tap and three in the base in readiness. The ornate, impressive cabinet cost $75 with "Japanned Trimmings," or $80.00 with silver-plate trimmings.[187]

The Evolution of Bottling

In 1788, George Washington wrote to his deputy quartermaster general, Colonel Clement Biddle, "I beg you will send me a gross of Mr. Hairs best bottled Porter if the price is not much enhanced by the copious droughts you took of it at the late Procession."[188] Hare (Hair) of Philadelphia was believed to be America's first porter brewer, and he was certainly bottling some of his brew by this time. By 1812, David Embree opened the first brewery in Cincinnati selling beer, porter, and ale, "put up in bottles and casks."[189]

Brande, in 1830, said beer must sit in the bottle six or eight hours before being corked, and then stand in "dry deal sawdust." Bottles were so expensive he recommended a bottle rack sufficient to hold all one had, and a double padlock to protect them. Empty bottles must be "constantly washed." Stale bottled beer could be made palatable by mixing one stale with two good, thus avoiding throwing it out."[190]

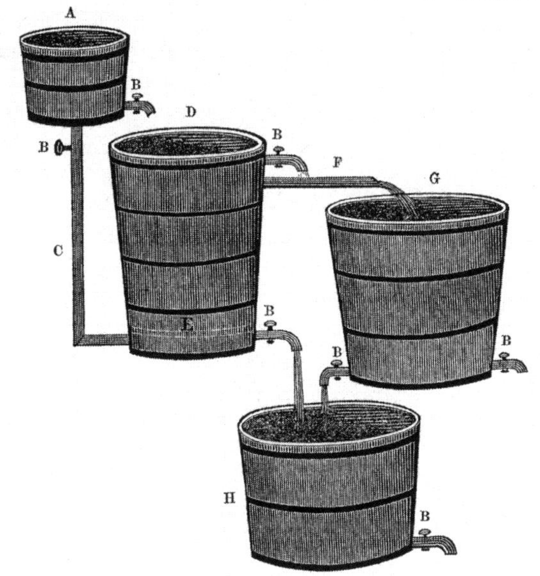

Fig. 17

A, the fountain.

B B, the cocks.

C, the trunk communicating with the space between the two bottoms.

D, the filtering-tub.

E, the false bottom.

F, the spout for carrying off the ascending liquor.

G, the receiver of the filtered liquor by ascent.

H, the receiver of the filtered liquor by descent.

Figure 1.11. Filtering system used for beer about to be bottled. (Byrn [1852] 2002, 58.)

The Bottling Process

Byrn's 1852 book had a lot to say about bottling beer.[191] It was potentially very profitable but could prove a heavy loss to the brewer who did not do it right. He recommended the beer, ale, or porter be filtered through sand and charcoal before bottling (fig. 1.11). "The corks should be steeped in strong spirit overnight before bottling. Once filled the bottles should be stored in dry straw in summer and sawdust in winter. If there is bubbling evident inside the bottle store them on end and pack with salt and they will not explode."

Bottling went from a completely manual job in 1870 to a totally automated job by 1900—at least in large breweries. A bottling plant once had a bottle washer, rinser, filler, corker, wirer, steam tank, labeling machine, and wrapping and packaging equipment.[192] The bottles to be shipped great distances were put into barrels padded with sawdust. Orders for a barrel of bottled beer are often seen in old brewery records. Wooden boxes later held a dozen bottles.

Filling a beer bottle was difficult because the beer could foam and leave space in the bottle. The oxygen in that space created a short shelf life. Special bottle-filling machines were developed in which a long spout went in the bottle to the bottom and filled it without foam. By 1891, some machines could fill up to twelve bottles at a time. Henes & Keller of Menominee, Michigan, sold a revolving filler with an air pump at that time.[193]

In the 1850s bottling tables were developed for the operator to insert the cork with foot or steam power.[194] In 1883, N. Wolf & Co. advertised for sale to breweries a corking machine for $35 and a steam powered version for $150.[195] The Crown Cork and Seal Company introduced automatic and semi-automatic bottling machines in 1898.[196] The patent-corker used to cork the bottles at the Billings, Montana, brewery in 1900 was a self-feeder and its primary virtue was saving labor. Visitors were impressed by it.[197]

For bottled beer, refrigeration apparatus cooled the beer to near freezing before the final filtering.[198] This helped prevent loss of carbonation and took out some albuminous substances that could cloud the beer.

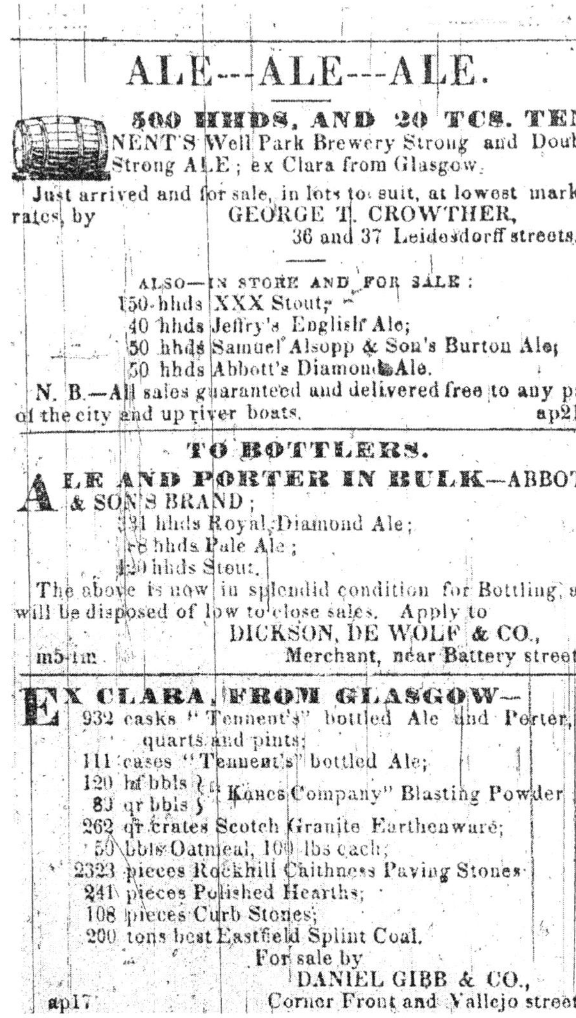
ALE---ALE---ALE.

500 HHDS, AND 20 TCS. TENNENT'S Well Park Brewery Strong and Doubl Strong ALE; ex Clara from Glasgow.

Just arrived and for sale, in lots to suit, at lowest marke rates, by GEORGE T. CROWTHER, 36 and 37 Leidesdorff streets.

ALSO—IN STORE AND FOR SALE:
150 hhds XXX Stout;
40 hhds Jeffry's English Ale;
50 hhds Samuel Alsopp & Son's Burton Ale;
50 hhds Abbott's Diamond Ale.

N. B.—All sales guaranteed and delivered free to any par of the city and up river boats. ap21

TO BOTTLERS.

ALE AND PORTER IN BULK—ABBOTT & SON'S BRAND;
[illegible] hhds Royal Diamond Ale;
[illegible] hhds Pale Ale;
[illegible] hhds Stout.

The above is now in splendid condition for Bottling, and will be disposed of low to close sales. Apply to DICKSON, DE WOLF & CO., Merchant, near Battery street. m5-1m

EX CLARA, FROM GLASGOW—
932 casks "Tennent's" bottled Ale and Porter, in quarts and pints;
111 cases "Tennent's" bottled Ale;
120 hf bbls } "Kanes Company" Blasting Powder
83 qr bbls }
262 qr crates Scotch Granite Earthenware;
50 bbls Oatmeal, 100 lbs each;
2323 pieces Rockhill Caithness Paving Stones;
241 pieces Polished Hearths;
108 pieces Curb Stones;
200 tons best Eastfield Splint Coal.

For sale by DANIEL GIBB & CO., Corner Front and Vallejo streets. ap17

Figure 1.12. Newspaper ad for imported beer in bulk.

Bottling by Subcontractors

Many brewers that shipped beer in the early days sent it in bulk and it was bottled at the destination. Some beer labels read "bottled at the brewery," which was often thought to be a superior site of origin. Local bottlers varied in their skill and diligence.[199]

Most, but not all, of the British malt liquors that were widely advertised and sold in the Mountain West and Pacific Northwest arrived in bulk at San Francisco and were then bottled there. Their advertisements appeared on the same page and column of a San Francisco newspaper in 1855[200] (fig. 1.12).

The Tennent's ales sold by Daniel Gibb & Co. were bottled, but the Tennent's Crowther sold was in hogsheads. Brewers Jeffrey, Samuel Abbott, and Alsopp were all shipping beer to the American west coast in bulk, and bottling firms were repackaging it.

Food and beverages in general might be imported in bulk or in small packages. European wines, for instance, were often transferred to English bottles in England because it was illegal to import small containers.[201]

Other Considerations of Bottling

Adolphus Busch was correctly credited with changing beer bottling from a relatively insignificant sideline into a major component of the brewing industry. Busch learned of Louis Pasteur's work before it was officially published, and decided, in a bold experiment, to incorporate the idea into a scheme to sell beer far and wide.

Several factors met simultaneously to make bottled beer a logical experiment for the company in the early 1870s. Busch had conquered the local market to as great an extent as seemed possible; further local growth would be slow and halting. Furthermore, railroads had reached from Atlantic to Pacific. And pasteurization could make bottled beer last long enough to reach distant markets.

The design of the original pasteurizer has not been found, but it must have involved several tanks, with the water heated and agitated by steam.[202] Two to five barrels of beer per hour could be so processed. New machines with great capacity were soon manufactured (fig. 1.13).

Figure 1.13. Pasteurizer for a medium-size brewery of 1909. (North American Wine and Spirit Journal 21, no. 6 (April, 1909): 32.)

The Cincinnati Chamber of Commerce, in its 1873–1874 report, explained the new situation for lager beer: "a new feature of the business is the bottling of this beer, which before the present year was substantially confined to kegs and barrels."[203]

Jos. Schlitz was bottling its beer in Cleveland as early as 1883.[204] In 1892, Detroit's Stroh's Brewery did the same, Pabst followed in 1896, and Anheuser-Busch had a major presence among bottled beers.

A pamphlet issued by the Bergner & Engel Brewing Company of Philadelphia about 1890 bore an excellent lithograph of the interior and operation of one of their bottling houses.[205] They shipped beer in half-barrel kegs to various depots, such as Richmond and Norfolk, Virginia, where it was bottled. Over fifty men, women, and youngsters toil in the picture. Two well-dressed men appear to be supervisors. On one side of the room, empty bottles are cleaned, then are moved to be individually filled, labeled, packed in boxes, and lifted out of the building on a primitive hoist. Eight half-barrels are suspended from the ceiling by chain hoists. One large barrel in the scene is labeled Tannhaeuser, 10 doz. Pints.

As late as 1891, the "New and Improved Goulding Bottle Washer" was for sale to breweries.[206] It apparently required an employee to put one bottle at a time on the end to be washed by a power driven brush that also sprayed cleaning solution inside. It seems excessively elaborate for the little specialized work that it does (fig. 1.14). Also available were bottle soaking tanks with a brush bottle washer and rinsing stand attached, or separate bottle rinsing machines could be purchased.[207] Other types and models washed up to six bottles at a time; a hand-powered machine washed two bottles at once.

A description of bottling at a small brewery before 1913 explains some of the complexities of the bottling process: The bottles were fed into a bottle washer machine where they were filled with a bottle cleaning solution. As the machine came around the solution emptied out of the bottles. They were dumped into a tray of water where they were rinsed, then placed in a machine which scrubbed and rinsed them inside. The bottles were filled with beer on a 12-spout hand-operated filler. When full a bottler put them on a stand where a crowning head came down and put a cap on the bottle. The filled bottles were put into a metal tray which held 24 bottles. These trays were stacked in an open cement-type 80-tray pasteurizer. The pasteurizer was filled with water until it covered the bottles, then steam-heated to a temperature of 140 degrees for 30 minutes, which pasteurized the beer. When the bottles cooled they were inspected. A machine put the label on the bottle to complete the process.[208]

Figure 1.14. Bottle soaker of 1909. (North American Wine and Spirit Journal 21, no. 6 (April, 1909): 32.)

Other types of pasteurization equipment existed. Tunnel equipment put bottles on a conveyor belt to be run through the pasteurizer that heated them to 140 degrees for about 20 minutes. In flash pasteurization the beer was run between two heat exchangers and heated very quickly to 160 degrees for about 20 seconds before being cooled.[209]

Labeling machines were developed such as those sold by The Bottlers' Specialty Mfg. Co., Chicago.[210] Some brewers wrapped each bottle in specially designed reusable paper wrappers to help prevent breakage. Bottles began to be packed in wooden boxes that held one, two, or four dozen bottles, and allowed the brewer to put advertising on the side.

As early as 1862, saloonkeepers were instructed in how to handle bottled beer. "Bottled beer should be kept in a cool place or a refrigerator, not in contact with the ice. The bottle ought to stand upright, so that any sediment will settle to the bottom. It is therefore, not advisable to pour the last dregs of the bottle into the glass."[211]

Archaeological investigations at the Klondike Brewery in Dawson, Canada, found side plate covers from a "Perfection Bottling Machine" within the back cellar depression. "This machine was patented in 1901 and could fill bottles with a variety of closure types."[212] It was manufactured in Chicago. In 1991 an elaborately decorated crown capping machine was reported at the site but it was no longer there when the archaeology was done.

The introduction of forklifts in the 1950s and 1960s reduced the physical labor of brewery workers. At that time full wooden cases of full bottles weighed 52 pounds, and later cardboard cases weighed 41 pounds. Half-barrels weighed 150 pounds.[213] Beer cans and metal kegs are discussed in the chapter on patents, trademarks, labels, and so on.

Ice and Refrigeration in Breweries

From Jamestown in the early 1600s, to Sir William Berkeley, governor of Virginia in 1665, to George Washington in the 1700s, to the ice wells excavated in Alexandria, Virginia, saving winter ice for use in the hot summer was an early American technological tradition[214] (fig. 1.15). In 1805, Boston merchant Frederick Tudor shipped ice from a pond in Lynn, Massachusetts, to Boston, the southern United States, the West Indies, and South America. By August, 1850, carts ran through San Francisco selling Boston lake ice at thirty cents per pound.[215]

The ice business began to flourish after 1850 when the demands of brewers, meat packers, and railroads (for cooling produce) caused the demand to mushroom. In 1874, ice consumption reached record levels in Cincinnati; ice sold for a record price of $12.00 per ton.[216] Despite competition from mechanical refrigeration devices, in 1888 brewers were the largest users of ice in America.[217] In fact, natural ice-cutting and artificial or mechanical ice production grew and prospered side by side for decades.

Breweries used 1,000 to 60,000 tons of ice per year depending on their size. The success of the breweries of this era depended in large part on the availability of large amounts of local ice, according to a number of brewing historians. Ensuring a consistent and timely supply of ice was difficult even in the north, and the logistical nightmare of warehousing and moving up to 60,000 tons of ice within the brewery icehouses, storage, and fermenting facilities created a desire for an improved system (fig. 1.16).

One Hundred Years of Brewing attributed most of the developments in the brewing industry from 1875 to 1900 to the use of mechanical refrigeration for many operations such as cooling the wort, fermentation, storage of beer and hops, racking, bottling, and malting.[218] Mechanical refrigeration reduced labor costs and saved space in the brewery. Breweries often ran independent businesses using this equipment as well, namely cold storage and ice-making for the retail sale of ice. The introduction of bottom-fermented beer in the 1840 to 1850 period further stimulated the need for ice.

A Revolution in Refrigeration Technology

Brewers therefore developed above-ground icehouses as well as systems that added salt to the ice to lower the temperature. In 1860, Carre patented his ice machine and inaugurated the modern brewery refrigeration era. David W. Davis developed and patented an ice and salt refrigeration system for breweries and first installed one in the Voigt brewery of Detroit in 1880[219] (fig. 1.17). Many devices were developed by many inventors, with various chemicals involved. All were based on the principle that a gas under pressure produces intense cold when that pressure is removed and rapid evaporation takes place. Not all brewery refrigeration

Figure 1.15. George Washington's icehouse at Mount Vernon. (Photo by author.)

Figure 1.16. An underground cellar with arched roof on which ice would be piled. (100 Years of Brewing, 1903, 146.)

Figure 1.17. D. W. Davis's Ice and Salt Refrigerator. Ice was piled on top, covered with salt, and the ice was replaced as needed, maintaining a constant cold temperature. (100 Years of Brewing, 1903, 122.)

systems were designed to make ice; many simply kept rooms or equipment cold or made cold water.

Franz Windhausen of Germany developed an ice machine first used in America in Cincinnati in 1875. Next, Daniel L. Holden of Kentucky developed an ice machine that was first used in 1877. Thomas L. Rankin developed a successful machine most notably used by the Jacob Rupert brewery of New York City. Also in 1877, David Boyle installed a 50-ton ice machine of his design at the Bemis & McAvoy brewery in Chicago.[220] The machine was an unparalleled success and Boyle sold twenty-two more machines in a period of only five months.[221] Theodore Krausch experimented with a system using both an ice machine and natural ice, and installed a prototype in the Ziegele's brewery in Buffalo about 1877.[222] In this device, an ammonia-based machine built by Silas Merchant of Cleveland had its piping running over natural ice.

Many other such developments in ice making and refrigeration emerged, so that a brewer in the late nineteenth century had many to choose from. Identifying an ice machine from its remains might prove difficult if no manufacturer names or patent numbers are located. Too many possibilities and variations exist for an archaeologist to be expected to identify a refrigeration machine on sight.

Ice cutting and mechanical ice machines coexisted for decades and evolved at different paces across the country. Breweries obtained ice through whatever means suited the location and economics of the individual plant. By 1864 a plant in New Orleans made 600 pounds of ice every three hours. By 1867 three ice plants were in operation in San Antonio, Texas.[223] In Idaho Territory an early mention of ice cutting is known from 1868 at Silver City high in the Owyhee Mountains.[224] Thirty-five years later, the sale of the Silver City brewery included the ice ponds formed above the city

Figure 1.18. Ruins of a brewery icehouse in Silver City, Idaho, which was once attached to the brewery, still stand in this ghost town at 6,000 feet of elevation. (Photo by author.)

by putting dams on Jordan Creek[225] (fig. 1.18). Little had changed in that isolated area. The Breckheimer brewery in Madison, Wisconsin, used a two-level icehouse with shafts to force the cold air down from the ice to the beer on the lower level.[226]

In 1886, Nick Kessler of Helena, Montana, installed a 15-ton Carbonic Acid Ice Machine built by Krupp Machine Works of Essen, Germany. It was powered by a 75-horsepower Hamilton Corliss engine. The machine broke down continually and exasperated all the local mechanics. Eventually, a German expert was brought over. An American-made machine later replaced it.[227]

Probably the first ice-making plant in Idaho was in Moscow in 1891, but the brewers went bankrupt and sold all their equipment in 1894. By 1903, only one manufacturer of ice was listed in the business directory for Idaho. In 1900, the most impressive room in the Billings, Montana, brewery for visitors was the cooling room with its "most complete [ice machine] in the West." It manufactured ice 6 inches thick across its pipes.[228]

In 1883, ice from northern Indiana sold for $7 a ton in Louisville. Soon, ice machines made ice at a cost of forty cents per ton.[229]

Many examples are known of the ice and brewery business intersection. The brewer of Carson City, Nevada, said this of his ice machines: "A 20-ton reciprocal Vulcan Ice Machine, driven by a 30-horsepower 2,300 volt electric motor, was installed to refrigerate the cellars and for the manufacture of ice. The ice plant had a capacity of five tons daily. The ice was delivered to local customers . . ." [230] Ice sales were a natural development for brewers who already cut tons each winter. One solution was to simply build a larger icehouse, fill it, and open a business delivering ice. Wagons and horses used for beer delivery were easily adapted to ice delivery. Whether ice originated in a machine or a frozen pond or river, the retail portion of the business did not change.

George Gans of Buena Vista Bar, Idaho, in 1866, owned the mining camp brewery with his brother Jacob, and a saloon with his brother Henry. George specialized in the ice portion of their businesses, advertising five cents per pound from his wagon, and orders could be left at the wagon or at Gans' saloon. Soon, he extended his sales territory to include Pioneer, Placerville, and Centerville, all within ten miles in the Boise Basin area. George and Henry dissolved their partnership in July 1866 with Henry staying in the saloon business and George still selling ice.[231] This mountainous region had an abundance of natural ice available for harvest.

From Natural Ice to Machine Ice to Refrigeration

Already in 1880, a British book on brewery construction opined "no brewery of any size should be without an ice-making machine, not necessarily for making ice, but for cooling water, it being more convenient to have a supply of cold water . . . than a supply of ice."[232] The author suggested that refrigerators ought to be selected based on simplicity of construction, ease of cleaning, and economy in the amount of water needed to cool a given quantity of wort.[233] Because the wort needed to be visible during cooling, refrigerators that were in a wooden case or "back" were unacceptable.

Edmund Jungenfeld built a series of new buildings for Anheuser Busch in 1878–79, including a new brew house, three vaulted cellars, four ice-storage houses, and a racking room that was cooled by overhead ice.[234] He advertised his expertise in building both icehouses and refrigeration machines. In 1882 this was all converted to mechanical refrigeration when three 50-ton Boyle machines—the largest such plant in the world at that time—were installed and these new buildings were converted.

In 1891 the A. Fitger & Company brewery of Duluth, Minnesota—a state abundantly gifted with natural ice—installed one of the first refrigeration machines in the region.[235]

The relationship between the use of winter-cut ice and machine-made ice is not at all uncomplicated. The example of the Pabst Brewery demonstrates increased production coupled with decreased use of natural ice. Milwaukee certainly had winter ice in abundance but refrigeration and increased production accompany each other.

According to the U. S. Census, only four ice-manufacturing plants existed in America by 1860, but by 1909, the number reached 2,004.[236] In 1880, the government estimated that only 15 square miles of ice 12 inches thick would supply the whole country.[237]

Cincinnati brewer Christian Moerlein revealed eight advantages of mechanical refrigeration over natural ice. He saved $8,000 to $10,000 per year in labor costs. Space devoted to ice was freed for other brewing functions, and dampness, which rotted timbers, was reduced. His machine cost $16,000 to $18,000 per year as opposed to the

Table 1.1. Pabst Brewery Expenditures for Natural and Machine-Made Ice, 1884–1892

Year	*Production, in barrels*	*Natural ice expenses*	*Ice machine expenses*
1884	375,983	$10,811	$33,468
1885	385,234	$14,312	$28,052
1886	436,438	$14,135	$39,505
1887	463,181	$527	$44,162
1888	505,638	$10	$43,263
1889	608,908	0	$43,919
1890	772,779	0	$49,866
1891	804,548	0	$64,366
1892	871,788	0	$55,691

Sources: Cochran, Pabst Brewing, 1948, p. 109. Stack, "Liquid Bread: An Examination of the American Brewing Industry, 1865–1940" (dissertation, University of Notre Dame, 1998), 90.

$60,000 spent on ice the previous year. Temperature control of refrigeration was more reliable and cleaning was easier. Refrigeration also provided two important production considerations: Year-round lager brewing was at last possible, and fermentation time was reduced from sixteen to eighteen days down to twelve to thirteen days.[238]

In 1882, Frederick W. Wolf of Chicago acquired the rights to the German Linde Ice Machine.[239] Wolf was an architect specializing in designing breweries, and more specifically, in designing the refrigeration systems for existing breweries. Later, the refrigeration systems were the major if not exclusive portion of the business. The drawings and blueprints from this company are housed at the Chicago Historical Society. This offers a major resource to identify and understand brewery refrigeration systems dating from the 1880s. In 1891, *The Western Brewer* opined, "almost every well appointed brewery is refrigerated by machinery, and nobody thinks of questioning their value."[240]

Thomas William O'Brien, owner of the Klondike Brewery in Dawson, Yukon Territory, installed an artificial refrigeration machine. The archaeologists believed this "a most curious addition to a building in the Yukon Territory where refrigeration is all but guaranteed for nine months of the year."[241] The efficiency that such a machine offers was demonstrated by this example.

Transportation

A brewery has great quantities of raw materials arriving and weighty products and by-products departing. An 1880 British brewery expert recommended that "arrangements should be made, where possible, to have a railway siding alongside,

Figure 1.19. Bartel brewery delivery wagons in 1909. (North American Wine and Spirits Journal 21, no. 6 (April, 1909): 32.)

or the brewery may be placed with advantage on the banks of a river or canal, whereby the expense of cartage may be saved."[242] The Lion Brewery of Cincinnati, for instance, was built in 1867 with one building facing the Miami and Erie Canal so it could use it for transportation.[243] Breweries in remote, mountainous areas of America, however, often had no hope of using any of these types of transportation.

The fine harbor in Milwaukee enabled the city's brewers to expand from shipping a mere 645 barrels in 1852 to 30,000 barrels in 1857, thereby establishing the city's reputation as a brewing center.[244]

Horses and Wagons

The Budweiser Clydesdale horses pulling their colorful beer wagon remain one of the most recognizable corporate symbols in American history. In the nineteenth century, beer deliveries were all made by horse-drawn wagons, and a brewery's sales territory was limited to the distance a team could travel, make the delivery, and return to the brewery in one day. A twenty-mile roundtrip from the brewery was about the maximum journey possible, and this distance was feasible only if the roads were exceptionally good (fig. 1.19).[245]

Horse-drawn wagons required space, maintenance, and labor. The essentials included a stable, tack and harnesses, storage for horse feed, and horseshoeing—which might be done on premises if the number of horses was sufficient. Veterinary bills were part of the equation, as was the removal of manure. Waterproof horse covers were purchased by some brewers to protect the horses in inclement weather.

Documentation from the Hamm's brewery revealed how important horses were:

> By 1886, thirty horses were needed to deliver Hamm's and by 1894 sixty magnificent Percheron horses found excellent quarters in the brewery stables. One hundred and fifty horses were an integral part of the brewery

by 1907, and beer was selling for seven cents a bottle. Some brewery wagons carried so much weight that they required three horses, and the drivers were paid an additional five cents an hour for driving the extra horse. The drivers usually started their day at six a.m., polishing the harness and equipment before loading at seven. Saturday was their busy day; deliveries often lasted until ten or eleven at night.[246]

Pabst Brewing Company claimed it once had eight hundred draft geldings to haul 250 to 300 beer wagons.[247] A small Illinois brewer explained in his journal that his two teams in late January of 1860 were hauling water and coal to the brewery, and delivering beer.[248] The number of horses in this example proved barely adequate, but he could afford no more.

Standard delivery wagons were manufactured for breweries. A catalog for the Empire Wagon Works of New York featured seven standard styles, including a Bottled Beer and Soda Wagon, and Quart Bottle Wagon. The twenty companies listed to which they had sold included Anheuser-Busch.[249]

Trains

When railroads spread to remote areas of the country, the brewers there quickly recognized the potential market this opened, and began to ship beer much farther than horses could carry it. Rail transport, however, had its own problems. Before the advent of refrigerated railcars, beer would become too hot in summer. In winter, beer shipments could freeze if the car got too cold. Empty kegs en route back to the brewery were sometimes damaged.

Railroads experimented with using ice to cool cargoes as early as the 1840s. "The essential feature of the refrigerated car, namely the end-ice bunker, had been developed by 1860 and was the accepted method of construction by 1870."[250] Well-insulated railroad cars had end bunkers that were packed with ice and the air was made to flow across the ice. Icehouses along the route replenished the ice. After World War II, mechanical refrigeration replaced this system so that frozen foods could be hauled.

The effect of the railroads on the expansion of the Pabst (originally Best and Company, named after founder Jacob Best and sons) Brewing Company's branch in St. Paul, Minnesota, was described in 1882:

> [the branch] was necessitated in facilitating the supply of the enormous demand coming from all quarters of the Northwest. At its inception only bottled beer was handled, but the trade increasing at such a rapid rate, it became urgently necessary in 1880, that bulk beer should also be handled. An apt illustration of the phenomenal advance made in consumption of their beer, will be found in the fact that at the date above mentioned [1875] only 120 cars were required; in 1880, this number had swelled to 210, and their sales this year [1882] will reach fully 350 cars, or more. Clearly speaking, their trade has increased from $12,000 the first year to $250,000 during the present. Their beer is handled in ½, ¼, 1/8 barrels, also in bottles by the case and cask. Fresh supplies of beer are received daily. The trade of this house extends throughout Minnesota, Dakota, Montana, and the Northwestern States and Territories, giving flattering indications of still further enlargement.[251]

Specific railroad connections could lead to unexpected beer distribution patterns. For instance, Dick & Bros. Quincy Brewery Co., of Quincy, Illinois, had a major presence in Kansas City due to the railroad bridge built across the Mississippi at Quincy in 1868, as well as the presence of the Hannibal Railroad Bridge across the Missouri completed at Kansas City in 1869.[252]

Trucks

In 1898, the Bohmische Brauhaus of Berlin, Germany, became the world's first brewery to use electric beer trucks.[253] In 1904, Anheuser-Busch became the first St. Louis brewery to acquire a fleet of trucks, not to replace, but to augment the horse-drawn wagons.[254] The initial fleet was a single gasoline-powered and twenty-seven electric-powered trucks from the Rainier Co. of New York City. The electrics went an impressive ten miles per hour and could traverse twenty-five miles on one charge. They could carry fifty full half-barrels or one hundred empties, or fifty cases of pint bottles.

An early truck company advertised that one truck replaced two team-drawn wagons, hence one less driver.[255] Also, the trucks did not care about inclement weather.

Breweries began to acquire delivery trucks about 1901 and horses began to decline in importance, but horses were as popular as trucks up until Prohibition. By the time it was repealed, the interim development of roads and trucks made the switch to motor vehicles complete.

A 1904 advertisement of the Electric Vehicle Company of Hartford, Conn., showed an Anheuser-Busch electric delivery vehicle.[256] Electric trucks were popular from 1900 to 1910; their chief defect was a short range before recharging. Comparing them exclusively to horses, their range seemed adequate. The Auto Car Equipment Company of Buffalo, New York, ran ads that proclaimed "Brewer's and Bottler's Truck, Capacity, Forty Cases, two electric motors, direct drive transmission, double chain drive, speed 8 miles per hour, endless solid rubber tires, powerful, noiseless, easily controlled."[257] An advertisement for General Vehicle Company of Long Island, New York, seemed concerned with the period of transition between transportation modes. "Prepare Now For July Weather. Get your order in for one or more G. V. Electrics before the warm weather so that the first hot week will not find you hiring trucks or exhausting your horses. Aside from its peculiar advantages for beer activities there is another reason this year for buying the Electric—the driver situation. Anyone can drive a G. V."[258] In 1914, this G. V. Company advertised that nearly eight hundred of their trucks were in use by brewers.[259] These companies no longer exist, however, and their electric trucks bore names now unfamiliar: American, Alco, Grabowsky, General Vehicle Company, and Columbia, for example.

By 1912 gasoline trucks were proving to be superior for brewery work.

> Mack Trucks, with their hard-rubber tires and cruising speed of 15 mph, started crowding the horses by the time of the First World War."[260] Anheuser-Busch once had a wagon-building shop to make beer wagons for itself and its distributors.[261] It changed to building truck bodies early in the twentieth century. A 1912 article revealed: "while the electric [truck] is holding its own on short radius hauls, it is entirely out of its class on long distances. Thus it happens that many breweries which started to experiment with small fleets of electrics and afterwards increased into big installations, have latterly begun to supplement their transportation equipment with gasoline wagons for outlying territories."[262]

Author Donald Wood uncovered a valuable description of the operation and maintenance of vehicles at Anheuser-Busch:

> The Anheuser-Busch motorized fleet at this time [1912] consisted of 24 passenger cars, 50 electric trucks, and ten gasoline trucks. Anheuser-Busch built a garage to house its fleet (the phrase the 1912 magazine used was 'accommodations for stabling the motor wagons'). The facility included a four-vehicle wash rack, a 10,000 gallon underground gasoline tank (large enough to permit buying in rail-tankcar-size lots), a machine shop with a four-ton overhead hoist, a paint shop, battery department, gasoline car repair pits, and an electric truck garage with receptacles for charging.[263]

Wood also found the forms drivers used to report vehicle problems. "On the vehicle maintenance forms used in the Anheuser-Busch garage, the electric truck driver had these categories for pointing out problems with the battery: burnt strap, burnt lug, burnt jumper, leaky jar, dead cell, burnt jar, battery needs filling, no power, runs slow, or clean jumper lugs. For electric motors, he could cite: armature burnt out, field coil burnt out, field coil grounded, motor hanger broke, clean commutators, turn commutators, new brushes needed, or new bearings needed."[264]

Gasoline trucks of these decades also had many unfamiliar names such as Marquette, Peerless, Hewitt, Kissel, Atterbury Model D, Dart, Kleiber, White, Dorris, Palmer-Moore, Horner, Pierce-Arrow, Schacht, Diamond-T, FWD, Sterling, Moreland, and the not so unfamiliar GMC.

During the 1920s the distribution of near beer seemed to employ ever more Ford trucks. American trucks were sometimes exported to other countries where they hauled foreign beer. Brewers might also own trucks such as horse haulers to transport draft horses around the country for promotional purposes.

Since Prohibition ended, the number of brands of trucks used in breweries has diminished, but the number of trucks has ever grown. Smaller breweries with smaller fleets of vehicles would nonetheless have the same problems to address. An archaeologist might uncover old vehicle parts or old vehicle maintenance equipment parts.

Miscellaneous Equipment and Supplies Found in Breweries

Up to this point, brewery equipment and processes have been considered on a piecemeal basis. As new technology emerged, brewers added one new device after another to their establishments. In some instances, a complete re-outfitting could happen all at once. In 1912, The Panhandle Brewing Company of Coeur d'Alene, Idaho took over the brewery of the bankrupt and liquidated Coeur d'Alene Brewing Company. Glaser and Schwarz company of Cincinnati removed all the old equipment and replaced it with a brewing outfit that had been removed from the Bavarian Brewing Company of Covington, Kentucky. The Bavarian Brewing Company had been redesigned with new equipment, so the entire old brewing operation and its gear made the journey to Idaho to begin a new life.[265]

Some of the equipment found around breweries, and perhaps found in the archaeological record, could come from a long list that might include: pumps for wort or beer; pressure-recording gauges with or without alarms; steam boiler and pump; grain scale; hop separator; hop press; thermometers; Kraeuse apparatus for regulating krausening; beer observer to see if beer is running clear into kegs; racking troughs; fire pitching machine; keg branding machine; cork washers; hand trucks for empty bottles; paper bottle wrappers; horse covers; brushes; manhole doors; bung hole bushings; de-alcoholizing machines to make near beer; and even robots that move barrels and cases.

If it is possible to sum up brewing equipment and development on the eve of Prohibition, an advertisement by Anheuser-Busch did just that. In 1912, it brewed 2.5 million barrels per year, malted 2 million bushels per year, and bottled 1 million bottles per day. Elevators stored 1.75 million bushels. The entire operation had 600,000 barrels lagering, produced 12,000 horse power at the steam plants and 4,000 horsepower at the electric plants, and kept its ice plants making 1,200 tons per day. The brewery used 325 tons of coal per day. About 50,000 train cars went in and out of their plants per year. The fleet included 1,500 refrigerated freight cars, 142 horses at the home plant, 78 wagons at the home plant, 74 auto trucks at the home plant, 483 horses at branches, 430 wagons at branches, and 47 auto trucks at branches. The company employed 6,000 people in St. Louis and another 1,500 at thirty-six branches.[266] An enormous number of artifacts were involved.

In contrast, F. B. Pastoor, brewer, sold his brewing equipment with house and lot in New York City in 1657 for 3,600 Guilders. The outfit consisted of an "under-trough, a cooler, a brew kettle, a wort tub, a barrel kettle and twenty half barrels and beer tubs, an iron vise, a little kettle, a wooden funnel, a brewer's sling, two gutters, a long and a short gutter, and a scoop."[267]

In 1974, the former equipment of the National Brewing Company of Detroit was sold at auction, and the items were advertised all over the country. The list was categorized into bottle shop equipment, brew house equipment, and engine room equipment. Other property included tanks, machine and maintenance equipment, forklift trucks, and a completely equipped laboratory. Included on the list, among many other items, were two eight-head cappers, a tray packer for six-packs, a Meyer Super Eye Empty Bottle Inspector, wort cooler, hop strainer, six-roll malt mill, 500-barrel copper brewing kettle, Bishopric 600 Bbl. Stainless Swirl Tank, fifty-one wood storage tanks, a Wittemann Gas Collecting System, and Enzinger Stainless Diatomaceous Filters.[268]

Comparing these lists illustrates the vast expansion of the brewing industry in terms of its material culture over the last three and one-half centuries in the United States.

NOTES

1. A vast number of fermented beverages were brewed by native peoples before the arrival of European colonists, but I have found no compelling evidence (either archaeologically or in the historical record) to suggest the influence of such practices on the development of brewing industries in the United States. The same is true of the brewing traditions brought to North America by enslaved Africans.
2. Patrick McGovern, an archaeological chemist at the University of Pennsylvania Museum, has done fascinating work detecting evidence of ancient fermented beverages by analyzing the residue on pottery shards. This has potential to be applied to historical-era artifacts related to beer.
3. F. W. Salem, *Beer, Its History and Its Economic Value as a National Beverage* (New York: Arno Press, 1972), 70. Reprint from c.1880.
4. Sanborn C. Brown, *Wines and Beers of Old New England: A How-To-Do-It History* (Hanover, New Hampshire: University Press of New England, 1978), xiii.
5. The quickest explanation of brewing reduced to its essentials that I have found is in Bill Owens, *How To Build a Small Brewery: Draft Beer in Ten Days*, 3rd ed. (Ann Arbor, MI: G. W. Kent, Inc., 1992). He discusses how to set up a brewery in a home with simple, inexpensive equipment such as a camping cooler to serve as a mash tun.
6. "The Story of Kessler Brewery, Established 1865 and How Beer is Brewed" (Helena, Montana, n.d., c.1952), 1.
7. Charles Edwin Dick, "A Geographical Analysis of the Development of the Brewing Industry of Minnesota" (dissertation, University of Minnesota, 1981), 125.
8. Herman Ronnenberg, "Beer for the Mining Frontier, Two Breweries Served Idaho's Famed Coeur d'Alenes," *American Breweriana Journal*, July-August 1990, 8.
9. Danny Morris and Jeff Johnson, *Richmond Beers, A Directory of the Breweries and Bottles of Richmond, Virginia*, 2nd ed. (Richmond, VA: 2000), 19.
10. William Brande, *The Town and Country Brewery Book; or Every Man his Own Brewer* (London: Dean and Munday, c.1830), 67–77.
11. George Scammel and Frederick Colyer, *Breweries and Maltings: Their Arrangement, Construction, Machinery, and Plant* (London: E. & F. N. Spon, 1880), 1.
12. Home brewers with high standards have been known to obtain brewing water from specific sources to suit a certain type of brew. Tap water would work for one kind of beer, but be less than ideal for another.
13. Editors of *The Western Brewer Magazine, One Hundred Years of Brewing* (Chicago: H. S. Rich & Co., 1903), 50.
14. Kevin Kious and Don Roussin, "The Breweries of Collinsville, Il.," *The Breweriana Collector* 132:10, 12.
15. Scammel and Colyer, *Breweries and Maltings: Their Arrangement, Construction, Machinery, and Plant*, 20.
16. *One Hundred Years of Brewing*, 1903, 51.
17. Apps, *Breweries of Wisconsin*, 32.
18. *Idaho Daily Statesman*, June 16, 1905, p. 3, c. 1–2. Brewery manager Rupert Maxgut sent water samples to Milwaukee for testing by Professor Ernest Hantke, the head of Hantke's Brewers' School and Laboratories of Milwaukee. Hantke's report was translated from German to English and published in the *Evening Capital*. He said Boise's water was clear, colorless, tasteless, and odorless; decidedly alkaline, with little if any sediment, little organic matter or sand, and only harmless bacteria present. A complete list was given of everything found in the water. He recommended gypsum be added at the brewery at the rate of 1 ounce per 10 barrels of water to make it hard enough to take on flavor and sweetening.
19. Dick, "A Geographical Analysis of the Development of the Brewing Industry of Minnesota," 67.
20. Dick, "A Geographical Analysis of the Development of the Brewing Industry of Minnesota," 68.

21. Burley and Will, "Special Brew: Industrial Archaeology and History of the Klondike Brewery," 42.
22. In 2007 Anheuser-Busch released "Redbridge," the first sorghum-based beer distributed nationwide. It was specially designed for people allergic to wheat or suffering from celiac disease. People with celiac cannot tolerate gluten, a protein found in wheat, barley, rye, oats, and spelt. See "A-B Takes Sorghum Beer National," *Real Beer Page Mail, The Free Monthly Beer News Digest for the Online Beer Enthusiast* (January, 2007), unpaginated, http://www.realbeer.com/news/articles/news-003063.php.
23. In May, 1935, Oskar Scholz, an Austrian native, arrived to take over as brewmaster in Idaho Falls, Idaho. Scholz said he was attempting to duplicate the original Pilsen formula and was using a ton of local honey each week to lighten the finished product. See *Post-Register* (Idaho Falls), May 3, 1935, p. 1, c. 3, and *Post-Register* (Idaho Falls), August 28, 1935, p. 7, c. 1–6. He left shortly thereafter to brew in Oklahoma.
24. Brande, *The Town and Country Brewery Book; or Every Man his Own Brewer*, ix.
25. "Know Your Malting Barley Varieties," *Brewers Digest* (May, 1985): 24–25.
26. M. L. Byrn, *The Complete Practical Brewer* (Philadelphia: Henry Carey Baird, 1852), 10.
27. Stack, "Liquid Bread: An Examination of the American Brewing Industry, 1865–1940," 71. Besides high nitrogen, there is also higher protein and thicker husk. These create problems for head retention and haze in the beer.
28. Peter H. Blum, *Brewed in Detroit: Breweries and Beers Since 1830* (Detroit: Wayne State University Press,1999), 31.
29. Byrn, *The Complete Practical Brewer*, 45.
30. Timothy J. Holian, *Over the Barrel: The Brewing History and Beer Culture of Cincinnati, 1800 to the Present. Volume One (1800–1919)* (St. Joseph, MO: Sudhaus Press, 2000), 12.
31. Susan Appel, "The Midwestern Brewery Before Prohibition: Development of an American Industrial Building Type" (dissertation, University of Illinois, 1990), 66.
32. Salem, *Beer, Its History and Its Economic Value as a National Beverage*, 102.
33. *Idaho Semi-Weekly World* (Idaho City, Idaho), November 5, 1875, p. 3, c. 2.
34. *Idaho Semi-Weekly World* (Idaho City, Idaho), November 5, 1875, p. 3, c. 2.
35. "The Story of Kessler Brewery, Established 1865 and How Beer is Brewed" (Helena, Montana, n.d., c.1952), 6.
36. *The Post Register* (Idaho Falls), August 29, 1935, p. 7, c. 4.
37. Bartell Nyberg, "For Coors, the Barley Must be the Best," *Lewiston Tribune* (Lewiston, Idaho), August 14, 1983, 2A.
38. *One Hundred Years of Brewing*, 1903, 53.
39. *One Hundred Years of Brewing*, 1903, 53.
40. *One Hundred Years of Brewing*, 1903, 53.
41. *One Hundred Years of Brewing*, 1903, 53.
42. Scammel and Colyer, *Breweries and Malting: Their Arrangement, Construction, Machinery, and Plant*, 59.
43. Herman Ronnenberg, *Beer and Brewing in the Inland Northwest* (Moscow, Idaho: University of Idaho, 1993), 54.
44. An excellent article on the botany of hops was in Gerard W. Ch. Lemmens, "The Breeding and Parentage of Hop Varieties," *Brewers Digest* 73, no. 5 (May 1998): 16–26.
45. Tomlan, *Tinged with Gold: Hop Culture in the United States*, 12.
46. I would recommend the American Hop Museum, 22 South "B" Street, Toppenish, WA, 98948, for its collection of hop growing machinery and general information on the extended subject.
47. Salem, *Beer, Its History and Its Economic Value as a National Beverage*, 101.
48. Stack, "Liquid Bread: An Examination of the American Brewing Industry, 1865–1940," 164.
49. In 1885 Indian hop-pickers who had been replaced by Chinese in the Seattle area attacked the Chinese, killing two and wounding four before peace was restored. See *Idaho Tri-Weekly Statesman* (Boise, Idaho), September 17, 1885, p. 1, c.3.
50. Kathleen E. Hudson Cooler, "Hop Agriculture in Oregon: The First Century" (master's thesis, Portland State University, 1986), p. 2.
51. Kathleen E. Hudson Cooler, "Hop Agriculture in Oregon: The First Century," 18.
52. Stack, "Liquid Bread: An Examination of the American Brewing Industry, 1865–1940," 72.
53. Stack, "Liquid Bread: An Examination of the American Brewing Industry, 1865–1940," 73.
54. "The Story of Kessler Brewery, Established 1865 and How Beer is Brewed" (Helena, Montana, n.d., c.1952), 6.
55. Brande, *The Town and Country Brewery Book; or Every Man his Own Brewer*, viii, ix.
56. *One Hundred Years of Brewing*, 1903, 57.
57. Helmut Kieninger, "The Influence of Raw Materials and Yeast on the Various Beer Types," *Brewers Digest* 58, no. 10 (Oct. 1983): 44. There are undocumented accounts of bottom fermented beers before 1810.
58. Stack, "Liquid Bread: An Examination of the American Brewing Industry, 1865–1940," 73–74.
59. *One Hundred Years of Brewing*, 1903, 97.
60. Arnold A. Millard, *History of the Carson Brewing Company* (n.p.: Brewery Press, 1980), 23.
61. *Lewiston Morning Tribune* (Lewiston, Idaho), July 4, 1906.
62. Kious and Roussin, "The Breweries of Collinsville, Il," *The Breweriana Collector* 133: 28.

63. Morris and Johnson, *Richmond Beers, A Directory of the Breweries and Bottles of Richmond, Virginia,* 209.
64. Will Anderson, *Beer, USA* (Hastings-on-Hudson, NY: Morgan & Morgan, 1986), 72.
65. Morris Weeks, Jr., *Beer and Brewing in America* (New York: United States Brewers Foundation, 1949), 52.
66. Cindy Renfrow, *A Sip Through Time: A Collection of Old Brewing Recipes* (published by author, 1995), 4, 9.
67. Brande, *The Town and Country Brewery Book; or Every Man his Own Brewer,* 154.
68. Byrn, *The Complete Practical Brewer,* 150.
69. Stack, "Liquid Bread: An Examination of the American Brewing Industry, 1865–1940," 168.
70. Millard, *History of the Carson Brewing Company,* 23.
71. Peter R. Guetig and Conrad D. Selle, *Louisville Breweries: A History of the Brewing Industry in Louisville, Kentucky, New Albany and Jeffersonville, Indiana* (n.p: Mark Skaggs Press, 1995), 50–51.
72. John P. Arnold and Frank Penman, *History of the Brewing Industry and Brewing Science in America* (Chicago: Self-published, 1933), 90.
73. Scammel and Colyer, *Breweries and Maltings: Their Arrangement, Construction, Machinery, and Plant,* 30–35.
74. Scammel and Colyer, *Breweries and Maltings: Their Arrangement, Construction, Machinery, and Plant,* 31.
75. Timothy J. Holian, *Over the Barrel: The Brewing History and Beer Culture of Cincinnati, 1800 to the Present. Volume One (1800–1919)* (St. Joseph, MO: Sudhaus Press, 2000), 121–122.
76. Holian, *Over the Barrel: The Brewing History and Beer Cuture of Cincinnati, 1800 to the Present. Volume One (1800–1919),* 122.
77. Holian, *Over the Barrel: The Brewing History and Beer Culture of Cincinnati, 1800 to the Present. Volume One (1800–1919),* 121.
78. Scammel and Colyer, *Breweries and Maltings: Their Arrangement, Construction, Machinery, and Plant,* 33.
79. Carl H. Miller, *Breweries of Cleveland* (Cleveland: Schnitzelbank Press, 1998), 82.
80. Burley and Will, "Special Brew: Industrial Archaeology and History of the Klondike Brewery," 46.
81. Burley and Will, "Special Brew: Industrial Archaeology and History of the Klondike Brewery," 51.
82. Ronnenberg, *Beer and Brewing in the Inland Northwest,* 75.
83. Scammel and Colyer, *Breweries and Maltings: Their Arrangement, Construction, Machinery, and Plant,* 50.
84. Holian, *Over the Barrel: The Brewing History and Beer Culture of Cincinnati, 1800 to the Present. Volume One (1800–1919),* 121.
85. Jerry Apps, *Breweries of Wisconsin* (Madison: University of Wisconsin Press, 1992), 153.
86. *100 Years of Brewing* (1903; repr. New York: Arno Press, 1974), 145.
87. Holian, *Over the Barrel: The Brewing History and Beer Culture of Cincinnati, 1800 to the Present. Volume One (1800–1919),* 123.
88. Byrn, *The Complete Practical Brewer,* 11.
89. *One Hundred Years of Brewing,* 1903, 58. Most of the malting explanation derives from this source.
90. John T. Flanagan, *Theodore Hamm in Minnesota: His Family and Brewery* (Minneapolis: Pogo Press, 1989), 14.
91. *One Hundred Years of Brewing,* 1903, 58.
92. *One Hundred Years of Brewing,* 1903, 59.
93. Ronald Jan Plavchan, "A History of Anheuser-Busch, 1852–1933" (dissertation, St. Louis University, 1969), 62.
94. Byrn, *The Complete Practical Brewer,* 9.
95. *One Hundred Years of Brewing,* 1903, 61.
96. Brande, *The Town and Country Brewery Book; or Every Man his Own Brewer,* 40.
97. *One Hundred Years of Brewing,* 1903, 67.
98. *One Hundred Years of Brewing,* 1903, 66.
99. *One Hundred Years of Brewing,* 1903, 66.
100. Plavchan, "A History of Anheuser-Busch, 1852–1933," 63.
101. Stack, "Liquid Bread: An Examination of the American Brewing Industry, 1865–1940," 76.
102. This is illustrated in *One Hundred Years of Brewing,* 1903, 67.
103. *One Hundred Years of Brewing,* 1903, 68.
104. Morris and Johnson, *Richmond Beers, A Directory of the Breweries and Bottles of Richmond, Virginia,* 2nd ed., 219.
105. Stack, "Liquid Bread: An Examination of the American Brewing Industry, 1865–1940," 92.
106. Brande, *The Town and Country Brewery Book; or Every Man his Own Brewer,* 53.
107. M. F. Janowitz, "Indian Corn and Dutch Pots: Seventeenth-Century Foodways in New Amsterdam/New York," in *Health, Sanitation and Foodways in Historical Archaeology,* vol. 27 no. 2 of *Historical Archaeology* (Joan H. Geismar and Meta F. Janowitz, eds.), 13.
108. Miller, *Breweries of Cleveland,* 22.
109. *Evening Capital News* (Boise, Idaho), October 25, 1907, p. 6, c. 2.
110. Holian, *Over the Barrel: The Brewing History and Beer Culture of Cincinnati, 1800 to the Present. Volume One (1800–1919),* 13.
111. Kious and Roussin, "The Breweries of Collinsville, Il.," 9.
112. "The Story of Kessler Brewery, Established 1865 and How Beer is Brewed" (Helena, Montana: n.d., c.1952), 4–5. In the 1950s, this mill and an ancient hand-operated mash tank pump were still on the brewery grounds.
113. Scammel and Colyer, *Breweries and Maltings: Their Arrangement, Construction, Machinery, and Plant,* 55.

114. *Ketchum Keystone* (Ketchum, Idaho), July 19, 1884, p. 3, c. 1.

115. Morris and Johnson, *Richmond Beers, A Directory of the Breweries and Bottles of Richmond, Virginia,* 2nd ed., 201.

116. Morris and Johnson, *Richmond Beers, A Directory of the Breweries and Bottles of Richmond, Virginia,* 2nd ed., 201.

117. *One Hundred Years of Brewing,* 1903, 77.

118. Scammel and Colyer, *Breweries and Maltings: Their Arrangement, Construction, Machinery, and Plant,* 61.

119. *One Hundred Years of Brewing,* 1903, 78.

120. *One Hundred Years of Brewing,* 1903, 78.

121. Peter H. Blum, *Brewed in Detroit: Breweries and Beers Since 1830* (Detroit: Wayne State University Press, 1999), 33.

122. Kious and Roussin, "The Breweries of Collinsville, Il.," 10.

123. Millard, *History of the Carson Brewing Company,* 23.

124. *One Hundred Years of Brewing,* 1903, 83.

125. *One Hundred Years of Brewing,* 1903, 83.

126. Kevin Kious and Don Roussin, "William J. Lemp Brewing Company," *American Breweriana Journal* 97 (March-April 1999): 5–13.

127. The kettle was saved for nostalgia and as a historical curiosity by later owners of the Lemp Brewery. It was mentioned in *100 Years of Brewing* as being kept by William J. Lemp, the owner in 1903. Descendant Edwin Lemp donated it to the Missouri Historical Society in 1955. From 1964 to 1977 it was on loan to the Falstaff museum in St. Louis but was returned when that institution closed.

128. Apps, *Breweries of Wisconsin,* 16.

129. *Real Beer Page Mail, The Free Monthly Beer News Digest for the Online Beer Enthusiast* (October, 1999): 11. According to this news clipping, this technique was revived in Tennessee in 1993.

130. During the late twentieth century, Stroh's Brewing of Detroit made a drive to become a major national brewer. The specialty it advertised was "fire brewed" beer. The company was still applying direct flame to the brew kettle and believed it produced a superior product.

131. *One Hundred Years of Brewing,* 1903, 89.

132. Scammel and Colyer, *Breweries and Maltings: Their Arrangement, Construction, Machinery, and Plant,* 79.

133. Morris and Johnson, *Richmond Beers, A Directory of the Breweries and Bottles of Richmond, Virginia,* 2nd ed., 202.

134. Kious and Roussin, "The Breweries of Collinsville, Il.," 21. The diary of the brewer and owner of this brewery, Henry Mayer, is a fantastic primary source for understanding the daily operations of a small midwestern brewery of the 1850s to 1860s.

135. Kious and Roussin, "The Breweries of Collinsville, Il.," 24.

136. *One Hundred Years of Brewing,* 1903, 89.

137. Millard, *History of the Carson Brewing Company,* 23.

138. Appel, "The Midwestern Brewery Before Prohibition: Development of an American Industrial Building Type," 65.

139. Byrn's 1852 book on brewing recommends the use of the instrument.

140. *One Hundred Years of Brewing,* 1903, 101.

141. In the mid-1990s, on a visit to the Hales Ales brewery in Spokane, Washington, I saw their open fermenters. These are the only contemporary examples I am personally familiar with. This brewery has since moved and no longer uses this system.

142. Stack, "Liquid Bread: An Examination of the American Brewing Industry, 1865–1940," 74.

143. Byrn, *The Complete Practical Brewer,* 42. This account uses common British brewing terms; a few decades later, terms derived from the German language were universally used by American brewers.

144. Apps, *Breweries of Wisconsin,* 35.

145. Apps, *Breweries of Wisconsin,* 160.

146. Morris and Johnson, *Richmond Beers, A Directory of the Breweries and Bottles of Richmond, Virginia,* 2nd ed., 206–207.

147. Burley and Will, "Special Brew: Industrial Archaeology and History of the Klondike Brewery," 99–100.

148. Millard, *History of the Carson Brewing Company,* 7.

149. Morris and Johnson, *Richmond Beers, A Directory of the Breweries and Bottles of Richmond, Virginia,* 2nd ed., 208.

150. Morris and Johnson, *Richmond Beers, A Directory of the Breweries and Bottles of Richmond, Virginia,* 2nd ed., 208.

151. Scammel and Colyer, *Breweries and Maltings: Their Arrangement, Construction, Machinery, and Plant,* 74.

152. Holian, *Over the Barrel: The Brewing History and Beer Culture of Cincinnati, 1800 to the Present. Volume One (1800–1919),* 129.

153. Holian, *Over the Barrel: The Brewing History and Beer Culture of Cincinnati, 1800 to the Present. Volume One (1800–1919),* 130.

154. Guetig and Selle, *Louisville Breweries: A History of the Brewing Industry in Louisville, Kentucky, New Albany and Jeffersonville, Indiana,* 24.

155. *Idaho Register* (Idaho Falls), October 29, 1887.

156. Guetig and Selle, *Louisville Breweries: A History of the Brewing Industry in Louisville, Kentucky, New Albany and Jeffersonville, Indiana,* 94.

157. Stack, "Liquid Bread: An Examination of the American Brewing Industry, 1865–1940," 151.

158. "The Story of Kessler Brewery, Established 1865 and How Beer is Brewed" (Helena, Montana, n.d., c.1952), 7.

159. Stack, "Liquid Bread: An Examination of the American Brewing Industry, 1865–1940," 138.

160. Scammel and Colyer, *Breweries and Maltings: Their Arrangement, Construction, Machinery, and Plant,* 100.

161. Scammel and Colyer, *Breweries and Maltings: Their Arrangement, Construction, Machinery, and Plant,* 101.
162. Kious and Roussin, "The Breweries of Collinsville, Il.," 9.
163. Morris and Johnson, *Richmond Beers, A Directory of the Breweries and Bottles of Richmond, Virginia,* 2nd ed., 213–214.
164. Morris and Johnson, *Richmond Beers, A Directory of the Breweries and Bottles of Richmond, Virginia,* 2nd ed., 215.
165. Kious and Roussin, "The Breweries of Collinsville, Il.," 23.
166. Apps, *Breweries of Wisconsin,* 79.
167. Apps, *Breweries of Wisconsin,* 79.
168. Holian, *Over the Barrel: The Brewing History and Beer Culture of Cincinnati, 1800 to the Present. Volume One (1800–1919),* 99.
169. Guetig and Selle, *Louisville Breweries: A History of the Brewing Industry in Louisville, Kentucky, New Albany and Jeffersonville, Indiana,* 34.
170. *One Hundred Years of Brewing,* 1903, 120.
171. Guetig and Selle, *Louisville Breweries: A History of the Brewing Industry in Louisville, Kentucky, New Albany and Jeffersonville, Indiana,* 33.
172. Olive R. Jones, "Commercial Foods, 1740–1820," *Historical Archaeology* 27, no. 2 (1993): 28.
173. Weisgerber papers, University of Idaho Library, Special Collections. Edith C. (Weisgerber) Taylor, Philip Weisgerber, and Marie E. (Weisgerber) White, *A Man, His Family, and His City* (n.p., self-published, 1982), 7.
174. Millard, *History of the Carson Brewing Company,* 33.
175. Morris and Johnson, *Richmond Beers, A Directory of the Breweries and Bottles of Richmond, Virginia,* 2nd ed., 216.
176. Morris and Johnson, *Richmond Beers, A Directory of the Breweries and Bottles of Richmond, Virginia,* 2nd ed., 212.
177. Holian, *Over the Barrel: The Brewing History and Beer Culture of Cincinnati, 1800 to the Present. Volume One (1800–1919),* 100.
178. Morris and Johnson, *Richmond Beers, A Directory of the Breweries and Bottles of Richmond, Virginia,* 2nd ed., 211.
179. *Idaho Tri-Weekly Statesman* (Boise, Idaho), June 5, 1886, p. 3, c. 1.
180. *Idaho Daily Statesman* (Boise, Idaho), March 23, 1890, p. 3, c. 3.
181. Del Vance, *Beer in the Beehive: A History of Brewing in Utah* (Great Salt Lake: Dream Garden Press, 2006), 16.
182. Kious and Roussin, "The Breweries of Collinsville, Il.," 23.
183. Ronnenberg, *Beer and Brewing in the Inland Northwest,* 74.
184. "The Story of Kessler Brewery, Established 1865 and How Beer is Brewed" (Helena, Montana, n.d., c.1952), 7.
185. Apps, *Breweries of Wisconsin,* 77.
186. Apps, *Breweries of Wisconsin,* 81.
187. Holian, *Over the Barrel: The Brewing History and Beer Culture of Cincinnati, 1800 to the Present. Volume One (1800–1919),* 189.
188. Bill Yenn, *The American Brewery: From Colonial Evolution to Microbrew Revolution* (St. Paul, Minn.: MBI, 2003), 17.
189. Holian, *Over the Barrel: The Brewing History and Beer Culture of Cincinnati, 1800 to the Present. Volume One (1800–1919),* 9.
190. Brande, *The Town and Country Brewery Book; or Every Man his Own Brewer,* 156.
191. Byrn, *The Complete Practical Brewer,* 186–188.
192. *One Hundred Years of Brewing,* 1903, 112.
193. Morris and Johnson, *Richmond Beers, A Directory of the Breweries and Bottles of Richmond, Virginia,* 2nd ed., 223, 235.
194. Deborah A. Hull-Walski and Frank L. Walski, "Brewing and Bottling in Harpers Ferry, West Virginia," Chapter 17 in Paul A. Shackel, ed., and Susan E. Winter, Principal Investigator, *Interdisciplinary Investigations of Domestic Life in Government Block B: Perspectives on Harpers Ferry's Armory and Commercial District,* Occasional Report No. 6, Regional Archaeology Program, National Capital Region, National Park Service, U.S. Department of the Interior, Washington, D.C., 1993, p. 17.5.
195. William L. Downard, *The Cincinnati Brewing Industry, A Social and Economic History* (Oberlin: Ohio University Press, 1973), 39.
196. Hull-Walski and Walski, "Brewing and Bottling in Harpers Ferry, West Virginia," p. 17.5.
197. David F. Slade, "History of the Billings Brewing Company" (thesis, University of Montana, 1971), 11.
198. *One Hundred Years of Brewing,* 1903, 107.
199. Beer bottles and closures are discussed in the chapter on Patents, Labels, and Containers.
200. *Daily Alta California* (San Francisco, CA), May 16, 1855, p. 2, c. 3.
201. Jones, "Commercial Foods, 1740–1820," 27.
202. Plavchan, "A History of Anheuser-Busch, 1852–1933," 69–71.
203. Downard, *The Cincinnati Brewing Industry, A Social and Economic History,* 36.
204. Miller, *Breweries of Cleveland,* 85.
205. Morris and Johnson, *Richmond Beers, A Directory of the Breweries and Bottles of Richmond, Virginia,* 2nd ed., 40–41.
206. Morris and Johnson, *Richmond Beers, A Directory of the Breweries and Bottles of Richmond, Virginia,* 2nd ed., 220.
207. Morris and Johnson, *Richmond Beers, A Directory of the Breweries and Bottles of Richmond, Virginia,* 2nd ed., 221.
208. Millard, *History of the Carson Brewing Company,* 33.
209. Stack, "Liquid Bread: An Examination of the American Brewing Industry, 1865–1940," 152.
210. Morris and Johnson, *Richmond Beers, A Directory of the Breweries and Bottles of Richmond, Virginia,* 2nd ed., 236.

211. Jerry Thomas, *The Bar-Tenders Guide, or How To Mix All Kinds of Plain and Fancy Drinks* (New York: Dick and Fitzgerald, 1862), 18.
212. Burley and Will, "Special Brew: Industrial Archaeology and History of the Klondike Brewery," 42. A drawing of this machine as it was advertised by the manufacturer is found on page 45 of this report.
213. Guetig and Selle, *Louisville Breweries: A History of the Brewing Industry in Louisville, Kentucky, New Albany and Jeffersonville, Indiana*, 25.
214. Herman Ronnenberg, "Idaho on the Rocks: The Ice Business in the Gem State," *Idaho Yesterdays* vol. 33, no. 4: 2.
215. Ronnenberg, "Idaho on the Rocks: The Ice Business in the Gem State," 3.
216. Holian, *Over the Barrel: The Brewing History and Beer Culture of Cincinnati, 1800 to the Present. Volume One (1800–1919)*, 127.
217. Ronnenberg, "Idaho on the Rocks: The Ice Business in the Gem State," 2.
218. *One Hundred Years of Brewing*, 1903, 121.
219. *One Hundred Years of Brewing*, 1903, 122.
220. *One Hundred Years of Brewing*, 1903, 121–134. The brewery was owned by Downer & Bemis according to Bob Skilnik, *The History of Beer and Brewing in Chicago, 1833–1978* (St. Paul: Pogo Press, 1999), 26.
221. Skilnik, *The History of Beer and Brewing in Chicago, 1833–1978*, 27.
222. Appel, "The Midwestern Brewery Before Prohibition: Development of an American Industrial Building Type," 206.
223. Ronnenberg, "Idaho on the Rocks: The Ice Business in the Gem State," 4.
224. Ronnenberg, "Idaho on the Rocks: The Ice Business in the Gem State," 5.
225. Ronnenberg, "Idaho on the Rocks: The Ice Business in the Gem State," 5.
226. Apps, *Breweries of Wisconsin*, 53.
227. "The Story of Kessler Brewery, Established 1865 and How Beer is Brewed," (Helena, Montana, n.d., c.1952), p. 6–7.
228. Slade, "History of the Billings Brewing Company," 10–11.
229. Guetig and Selle, *Louisville Breweries: A History of the Brewing Industry in Louisville, Kentucky, New Albany and Jeffersonville, Indiana*, 57.
230. Millard, *History of the Carson Brewing Company*, 23.
231. *Idaho World* (Idaho City), June 9, 1866; July 7, 1866; August 4, 1866.
232. Scammel and Colyer, *Breweries and Maltings: Their Arrangement, Construction, Machinery, and Plant*, 86.
233. Scammel and Colyer, *Breweries and Maltings: Their Arrangement, Construction, Machinery, and Plant*, 85.
234. Appel, "The Midwestern Brewery Before Prohibition: Development of an American Industrial Building Type," 210.
235. Dick, "A Geographical Analysis of the Development of the Brewing Industry of Minnesota," 105–106.
236. Ronnenberg, "Idaho on the Rocks: The Ice Business in the Gem State," 6.
237. Ronnenberg, "Idaho on the Rocks: The Ice Business in the Gem State," 7.
238. Holian, *Over the Barrel: The Brewing History and Beer Culture of Cincinnati, 1800 to the Present. Volume One (1800–1919)*, 127–128.
239. Susan Appel, "New Finds in Brewery Architecture," *The Breweriana Collector*, Spring 1999, 7.
240. Appel, "The Midwestern Brewery Before Prohibition: Development of an American Industrial Building Type," 206.
241. Burley and Will, "Special Brew: Industrial Archaeology and History of the Klondike Brewery," 89.
242. Scammel and Colyer, *Breweries and Maltings: Their Arrangement, Construction, Machinery, and Plant*, 2.
243. Appel, "The Midwestern Brewery Before Prohibition: Development of an American Industrial Building Type," 126.
244. Apps, *Breweries of Wisconsin*, 8–19.
245. Dick, "A Geographical Analysis of the Development of the Brewing Industry of Minnesota," 84.
246. Quoted in Dick, "A Geographical Analysis of the Development of the Brewing Industry of Minnesota," 23–31.
247. Donald F. Wood, *American Beer Trucks* (Osceola, WI: MBI Publishing, 1999), 8.
248. Kious and Roussin, "The Breweries of Collinsville, Il.," 11.
249. This ad is undated but mentions "Telephone Connection," which would place it about 1890. Morris and Johnson, *Richmond Beers, A Directory of the Breweries and Bottles of Richmond, Virginia*, 2nd ed., 137.
250. H. James Maxwell and Bob Sullivan, Jr., *Hometown Beer: A History of Kansas City's Breweries*, (Kansas City: Omega Innovative Marketing, 1999), 75. Railroad sidings at breweries are covered in the chapter on brewery architecture.
251. Quoted in Dick, "A Geographical Analysis of the Development of the Brewing Industry of Minnesota," 230–231.
252. Maxwell and Sullivan, *Hometown Beer: A History of Kansas City's Breweries*, 5.
253. Plavchan, "A History of Anheuser-Busch, 1852–1933," 63.
254. Plavchan, "A History of Anheuser-Busch, 1852–1933," 63.
255. Holian, *Over the Barrel: The Brewing History and Beer Culture of Cincinnati, 1800 to the Present. Volume One (1800–1919)*, 124.
256. Wood, *American Beer Trucks*, 8.
257. Wood, *American Beer Trucks*, 10.
258. Michael C. Hennech and Doug Pierson, *Beer Trucks, A Photographic Salute, 1870–1992* (Texas: Ale Publishing Co., 1992), 12.

259. Hennech and Pierson, *Beer Trucks, A Photographic Salute, 1870–1992,* 13.

260. Dick, "A Geographical Analysis of the Development of the Brewing Industry of Minnesota," 232.

261. Wood, *American Beer Trucks,* 13.

262. Wood, *American Beer Trucks,* 13–14.

263. Wood, *American Beer Trucks,* 14.

264. Wood, *American Beer Trucks,* 14.

265. *The Western Brewer* 37, no. 6 (June, 1912): 286.

266. Holian, *Over the Barrel: The Brewing History and Beer Culture of Cincinnati, 1800 to the Present. Volume One (1800–1919),* 213.

267. Arnold and Penman, *History of the Brewing Industry and Brewing Science in America,* 48.

268. Classified Ad, *Los Angeles Times* (Los Angeles, CA), February 24, 1974, part II, p. 4.

CHAPTER TWO

Brewery Architectural Developments: 1500–2000

In older eastern cities, breweries that were built during the colonial period were often of superior construction, though not of special design, and their function as a brewery was not obvious to the eye. As brewing moved west with the population, brewery construction tended to be similar to that of the other buildings in a given settlement. Small breweries were located in frame houses, barns, and even log structures. Most breweries had an interior design based on gravity flow, so the water, wort, or beer did not have to be pumped to desired levels with hand pumps. A slight drop was engineered between each level, and therefore breweries tended to take up more horizontal than vertical space.

Buildings such as these, that blend into their surroundings, are called vernacular buildings. They are built according to local custom, not professional engineering or architectural concepts.[1] In the case of early breweries, there were few specific architectural features. One of the only common specialized features was a louvered cupola on the top for ventilation, a necessity to compensate for the steam created by open kettles, which condensed and dripped all over. Early breweries were also sometimes underdesigned for their function. Rickety stairs and weak floors—sometimes lethally dangerous—were common.[2]

A brewing facility built in 1748 in Burlington, New Jersey, by Thomas Shaw featured a brewery, malt house, and mill.[3] It incorporated a new invention—copper coolers—that held cold water and were dipped into the brew kettle to cool the wort after boiling. Also on the premises were millstones, tuns, and a kiln. The George Painter brewery in Burlington in 1800 measured 91 feet by 36 feet, with a malt house and its own dock on the Delaware River. Both breweries were probably made of wood and would have manufactured top-fermented ales.

An undated drawing exists of Anthony Morris's second brew house, erected about 1745, at Dock and Pear Streets in Philadelphia.[4] It shows a close-set group of brick buildings, ranging from three to five stories high, with "no particular stylistic pretensions."[5] The Globe Brewery of Baltimore, dating from 1744, had three excellent underground brick vaults that measured 120 feet long, 75 feet wide, and 9 feet high, and drew extravagant praise still in 1903.[6] The above-ground brewery was a single story.

According to brewery architectural historian Susan Appel, the period from 1810 to 1860 in which midwestern breweries moved from primitive to sophisticated was poorly recorded and the chronology of change has remained difficult to pin down.[7] Archaeology could add greatly to our knowledge of this.

Jacob Salmon, a flat-boatman from Virginia, moved to Indiana in 1823 and constructed one of the first breweries in the "West." This was noteworthy because a written description of the plant he built survived. A drawing of the buildings accompanied this source.

> [It was] a rough, low one-story stone structure. . . . about thirty feet wide and sixty feet long, the stone used in its construction being large square blocks, and the walls were about 30 inches thick. It was located on the abrupt bank of a deep ravine, making the rear two stories in height, and here were located the storage vaults. These were three in number, about nine feet wide, extending under the building a distance of about twenty-five feet. The stone building was used in part for a family residence. The log structure . . . was wider than the stone one, about the same length, built in primitive style, and contained the cooling floor, mashing tubs and malting floor, the upper story being used for the storage of grain. Water for the use of the brewery was obtained from a spring in the hillside above the road, and was conveyed by a v-shaped trough, supported on up right posts set at regular intervals.[8]

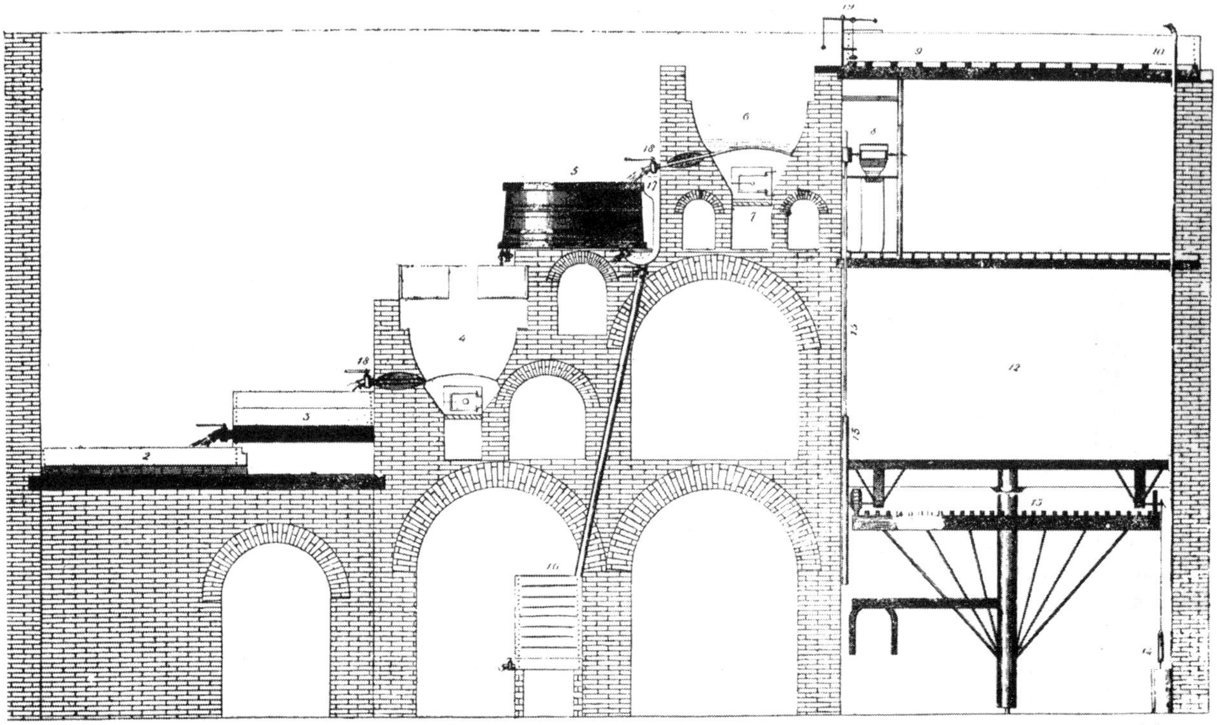

Figure 2.1. John Levesque's plans for a brewery, published in 1853. (100 Years of Brewing, 1903, 17.)

One of the first breweries in St. Louis, Philipson & Habb's, was a two-story frame structure.[9] The earliest breweries in Cincinnati were also frame buildings. One example from that Ohio River city from 1855 gave a great deal of information.[10] On the insurance maps, the thirty-year-old Walker Brewery had become a rectangular complex of several structures around a narrow open passage. There was a steam boiler and then two buildings on either side of it, which were considered at special risk for fire.

As modern equipment was added to existing breweries on a piecemeal basis, new additions or alterations to the buildings took place. "The large ale breweries at that time were of substantial and solid construction, though little importance was, as a rule, attached to their outward appearance."[11]

The Lemp Brewery of St. Louis was carefully analyzed in a dissertation on brewery architecture. Adam Lemp's first brew house—and home—was a small stone structure erected on Second Street, between Walnut and Elm, in 1841.[12] The brewery was rectangular, two stories high, with a simple chimney. A gabled roof held a louvered circular opening. A large round-arched entry on the ground floor stood below the gable, with two round arched windows on the second floor, two more in the gable wall, and three on the long side of the building. There was a one-story addition with a round-arched entry on the short end and a plain long side. Dr. Appel felt the round arched forms and "what could be called an oculus window in the gable seem to look forward vaguely to some Romanesque stylistic features of the 1860s lager breweries."[13] She concluded, however, these were simply small, vernacular structures, but substantial and built to last.

Early in the Lemp brewery's history, a two-story stone cellar was added. It was four bays wide and two deep, with shuttered round arched windows, quoined corners, and a hipped roof. On a one-story porch in the center of a long side was the entrance. A masonry pit in front was probably the entrance to the subterranean storage space. Shortly after this addition, Lemp began modifying the cave he would later use when the brewery location changed.

Londoner John Levesque, in 1853, published a design for a modern ale brewery of that period.[14] It was made of brick with arches to support the roof. The hot water started on the third floor, moving downward as the brewing process progressed. A mill track or horse wheel provided power. A plodding horse seemed a primitive feature in a design that otherwise appeared relatively modern (fig. 2.1).

With the advent of lager breweries in America, architecture had to accommodate different needs. The longer, colder storage phase made the development of cold storage necessary. At first, brewers brewed in cold weather and used naturally cold areas such as caves or underground cellars to lager

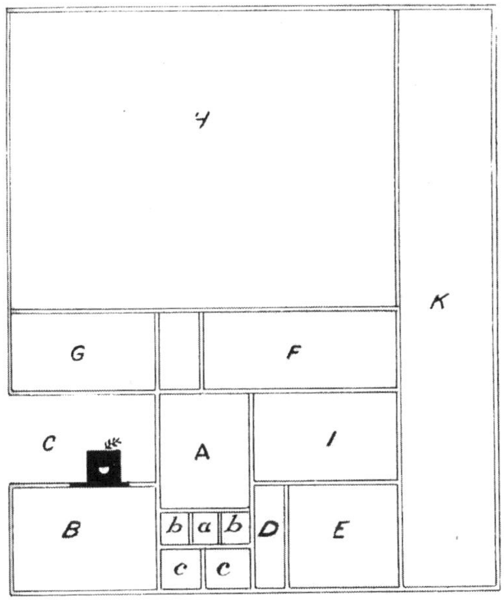

GROUND PLAN, MODEL BREWERY.

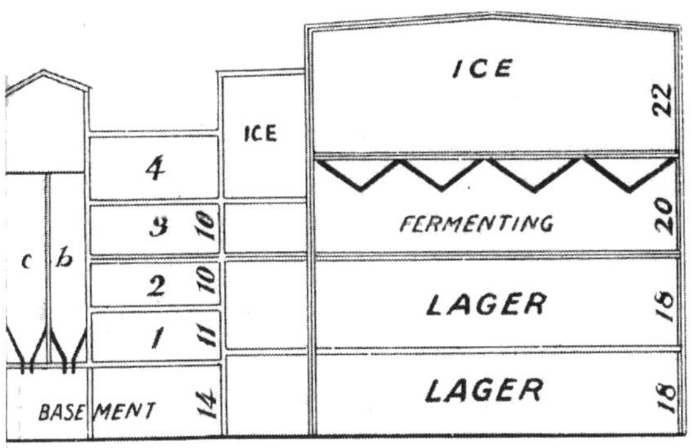

Figure 2.2. The design of a brew house in 1853. Notice more gravity flow than in earlier structures, and brick arches used in the construction. (100 Years of Brewing, 1903, 136.)

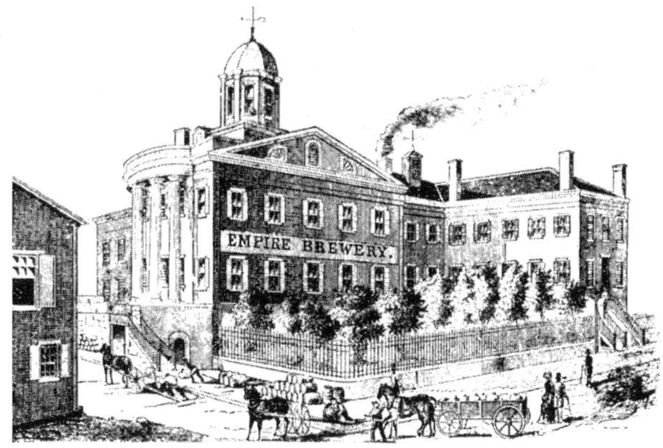

Figure 2.3. The Empire Brewery of New York City in 1846. The tall columns and cupola indicate interest in architectural features. (100 Years of Brewing, 1903, 197.)

(age) the beer. Then, as mechanical refrigeration became available, above-ground stock houses became the rule. As with all equipment and architectural developments in the brewing industry, some brewers changed quickly and others kept old ways for decades (fig. 2.2).

From 1810 to 1860, midwestern breweries went from primitive to sophisticated in a process of development that has been poorly recorded.[15] The early breweries sometimes served as foundations for later ones and helped determine how later plants would be sited or arranged. Successful breweries tended to become a series of expanded and new buildings until their land was filled with structures. While original buildings might have been frame or stone, brick eventually became the preferred material for breweries (fig. 2.3). The switch from ale to lager resulted in the construction of extensive underground cellars. Breweries always had fires for heat and consequently required chimneys, but the advent of steam power led to the characteristic giant brewery chimneys designed to carry enormous amounts of smoke skyward.

A German brewer who toured America in 1867 described the first-class lager breweries he observed as equal to the large breweries of Munich or the second-class breweries of Vienna.[16] The breweries in small towns that he saw were primitive and small in the extreme. Within a few years, these places were no longer competitive and were forced to close. By 1880 an American brewer, Andrew E. Leicht, wrote of his experience visiting the Anton Dreher brewery at Schwechat, near Vienna, Austria, in 1880. He found it large but not handsome, with buildings scattered around the grounds without plan. The brew house was too small and no icehouse was present. He felt that it trailed, not led, American breweries in innovation.

The Western Brewer, also in 1880, compared the Culmacher Brewery of Bavaria with American breweries and opined, "the buildings and plant [of Culmbacher] do not compare favorably"[17] (fig. 2.4). American breweries had experienced two decades of physical growth and development. The Civil War was often credited with spreading the belief that modern industry could perform miracles in the quantity and value of their products. The rest of the nineeenth century was the perfect time and atmosphere for the American brewing industry to mushroom and demonstrate the truth of that belief.

The Farmington Brewery, in Fillmore, Washington County, Wisconsin, was built in the 1860s and closed in the 1880s. "The building's construction followed the Old German half-timber architectural style."[18] This illustrates that neither frame nor brick construction were universal.

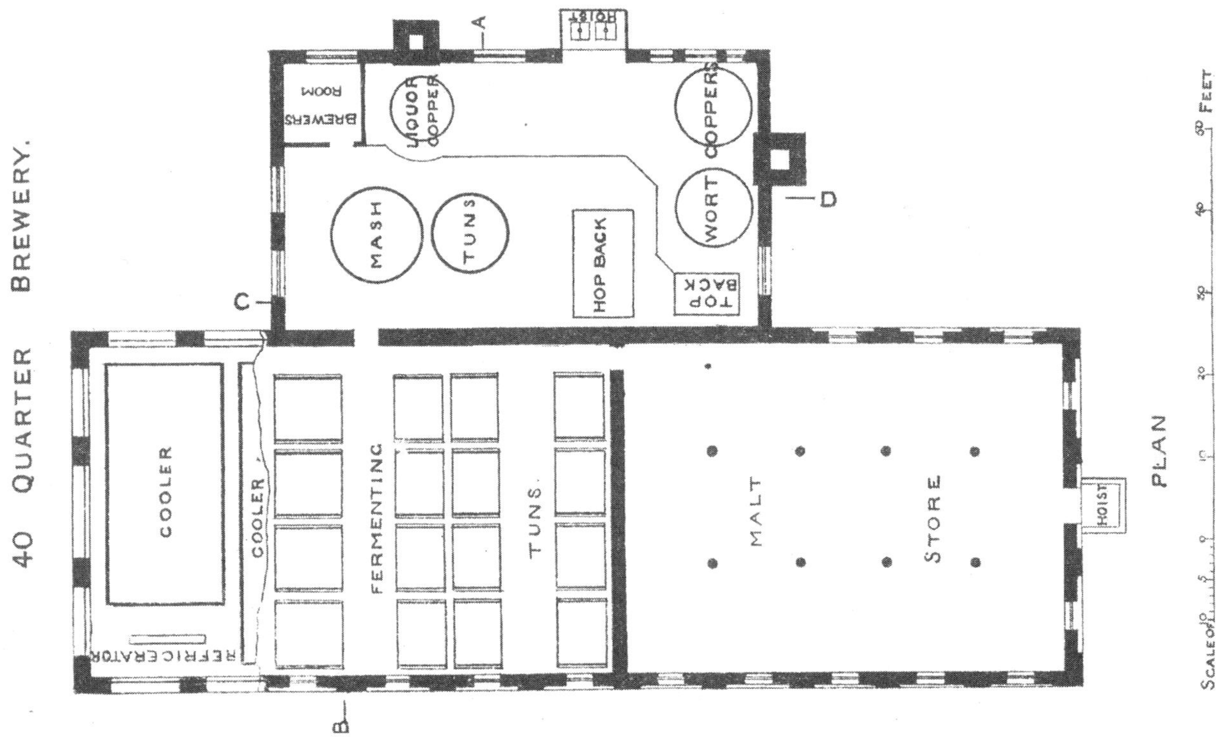

Figure 2.4. A German brewery design of 1880. (Colyer, 1880, plate 8.)

Emergence of Architecture Based on Function with Consideration to Style

In the 1860s the great increase in beer production and sales, the growth of lager beer as a percentage of the total malt liquor produced, the Romanesque Revival architecture trend that had first reached America in the 1840s, and perhaps other unidentified factors led to the building of breweries in a Romanesque style. Three lager breweries built in Cincinnati in 1863, 1866, and 1868 "bore a definite physical resemblance to one another and differed remarkably from pre-Civil War breweries; they consciously adopted a Romanesque Revival style that signaled a change of attitude in design of industrial buildings after the Civil War."[19]

Among the recognizable features of these trend-setting, brick construction breweries were "round-arched and oculus windows aligned in regular patterns and set in pilastered walls capped with decorative corbelling and hints of battlements."[20] They also had smooth, solid-looking walls and visual balance between the numerous windows. Other Cincinnati breweries showed similar stylistic features during this period. The Dayton Street Brewery went even further, being more ornate, and the "bracketing under the eaves here also shows how Romanesque and Italianate qualities could mingle in this period."[21]

The Romanesque style was versatile and the most noticeable feature was the round arch. Generally, the style was recognizable from the exterior, but the interior had no noticeable structural form. The highly recognizable exterior tended to advertise the business within and demonstrate to the world that businessmen cared about beauty.

Some architectural historians do not call this style Romanesque in its American manifestation, but believe it derived from the German *Rundbogenstil*, or "round arched style."[22] American architects were greatly influenced by German styles from about the 1840s; these stylistic components were found in other industrial buildings of the period. Even in Germany, lager brewing was not very old, but it was spreading and influencing architecture as specialized breweries were built to accommodate the new beer style.

Appel cited the Philadelphia breweries of Bentz & Reyle, built in 1866, and the Gustavus Bergner Brewery, begun in 1857, as contemporary examples that did not follow the new stylistic ideas.

Appel concluded, "The 1860s breweries in the *Rundbogenstil* express something of an ideal moment in the history of this building type in America. Compared with their vernacular predecessors, they were grand in scale, functionally more sophisticated, and intentionally well-designed, firmly constructed buildings. And while possessing an attractive and recognizable style, they also maintained a simple modesty that prevented them from becoming too powerful in their visual impact. Where their horizontal breadth filled space and stated effectively that the businesses housed in these structures were significant, neither their vertical scale nor their decorative trim was overwhelming. They expressed modernity and pride, ethnic and otherwise, but they also fit in, as the immigrant brewers who built them did in their new homeland."[23]

Architect Fred Baumann published an article in the November, 1876 issue of *The Western Brewer* along with a drawing and description of a model brewery for that period:[24]

> In these plans c and b are the bins for the storage of malt and grain, of which there should be two for each, in order that one bin may be wholly emptied and cleaned before touching the other, for the bottom contents of a bin are the last to flow through the pans. The elevator within handles the stuff and puts it through the crushing mill, placed on the fourth floor, whence it flows into the mash tub, placed on the third floor. Hot and cold water are let into the mash tub from tanks placed on the floor above. The mash juice is let into the seed-kettle, placed on the second floor, wherein the wort is prepared in whatever manner preferred by the brewer. The boiled wort flows into a large settling tank, placed on the first floor, when it is let over a Baudelot cooler, placed in the basement.
>
> The basement story further contains the engine, from two to four pumps, an air-pump for the discharge of the beer from the larger vats, and a suction fan for ventilating the fermenting room. From the Baudelot the wort is directly pumped into a vessel placed over the fermenting vats and under the ceiling of the room, wherein it is again allowed to settle before being discharged onto the fermenting vats.

Two points it is necessary to observe in a good brewery, though often neglected. First, no vapor should be allowed to escape within the building. Second, the fermenting room should be properly ventilated. Cool, fresh air is in all cases supplied from cold-air chambers overhead, which rapidly descends to the floor.

Frederick Colyer's British book on brewery construction of 1880 suggested the design selected allow for expansion, and each part of the brewing process be lower than the previous one to use gravity and therefore require the liquid to be pumped only once in the whole process.[25]

Steam engines or power plants greatly influenced brewery architecture as well. In some breweries, the boiler was inside the building against an outer wall. By 1875, safety concerns led the Lemp Brewery in St. Louis to place its boilers "set back underground in a cellar separately built for them apart from the main building."[26] By 1879, similar separation of the boilers from the main building was found in two other St. Louis breweries, the Samuel Wainwright & Co. Brewery and the Anheuser-Busch brewery.

Transitions in Building Materials

After the mid-1870s, attention was paid to fireproofing breweries and malt houses. Construction in brick and stone continued but iron, and later steel, framing was added, along with cement covered brick floors on iron columns.[27] Some parts of breweries were covered in iron, roofs were sometimes patent sheet iron, and stands and supports for equipment were iron. Hollow tile, glazed tile, and cement foundations became popular. By the 1890s, asphalt floors were common. Developments in industrial construction in general were also found in breweries of the period (fig. 2.5). As Dr. Appel summarized: "Together, these materials and methods brought about new kinds of spatial settings for the production of beer that were more open and easily supervised, as well as cleaner and safer than ever before."[28]

For example, the Windisch-Mulhauser Brewing Company of Cincinnati built a new brew house in 1888 of stone, glass, and iron that measured 197 feet long and 150 feet wide, and stood five to six stories high.[29] It had a bowed mansard roof capped with three large cupolas, and the iron interior allowed for many windows. The look was more vertical than horizontal, with small turrets and tall exhaust pipes.

Not all innovations in brewery architecture were a step forward. One development was rather poorly conceived. "By the mid-1880s the enormous brew kettles were often stabilized with the interlacing support and structural features of the towers themselves. This dependent support technique probably explains why almost all of the brew house towers of the remaining brewery plants found throughout the city [Chicago] today have been removed."[30] Once the kettles were removed the walls could have collapsed unexpectedly.

After 1900, new breweries were not constructed with wooden floors, stairs, or beams. Iron, stone, brick, and cement were the preferred materials. After a remodeling of the Lackman Brewery of Cincinnati a description said: "In this building, except one small house, there is not an inch of wood. Stone, iron, steel and brass are its components. It is fireproof, enduring, like the memory of the good man who planned it, whose four sons carried out his designs."[31]

Around 1900—with the first known application in Newport, Kentucky, in 1888—electricity began to replace steam power in breweries.[32] The cumbersome shafting, pulleys, and belts of the steam age were replaced with relatively small electrical engines serviced by wiring that required little space. Electrical engines also produced very little heat in comparison. Breweries were no longer built to accommodate the transmission of steam power.

Figure 2.5. Lauer's Brewery in 1880. (Salem [1880] 1972, 74–75.)

Breweries: A New Category of Architectural Style Arises

Architects Wolf and Lehle of Chicago designed the Foss-Schneider Brewery in Cincinnati in the mid-1880s so that it was a "striking architectural element of the city" [33] and represented a transitional design for brewery architecture of the period:

> If it resembled the general lines of earlier breweries, Wolf and Lehle's design for Foss-Schneider was also far more decorative than its predecessors. Its street façade was cohesive but taller and much more varied, in line with the eclectic nature of American architecture of this period. Arranged in five principal if irregular bays, the brewery was apparently four stories high, but crowned with a tall bowed mansard roof that made room for a fifth floor. Its entire exterior was a rich mixture of brick, stone and terra cotta, lively and polychromatic, with a High Victorian flavor of mixed heritage. There were both gothic and Romanesque windows, some with Venetian banding, and others with machicolated or flat-topped frames. Stepped and steep-sided gables suggested Flemish or German sources, and French Baroque or Second Empire overtones appeared in the roof form and its circular louvered dormers. Nowhere did this design show the balance between wall surfaces and openings, nor the simplicity of 1860s breweries. The wall was no longer a smooth foil for orderly and largely repetitive openings, but far more plastic, colorful and open. If each bay was still carefully ordered by aligning its windows and doors in both directions, variety was clearly more important than regularity to Fred Wolf and Louis Lehle.[34]

Architect Charles Stoll of Brooklyn designed the John F. Wiessner Brewery in Baltimore in the late 1880s. It "remains probably the most extreme example of brewery eclecticism ever constructed in the United States."[35] No label seems to properly fit it. Some have tried "Teutonic Breweryesque" or "a concoction of German Romanesque, French Second Empire and Italianate designs."[36]

In 1912, the Bay City Brewing Company of San Diego had Richard Griesser design its new brewery in the Mission style, which was considered perfect for San Diego but unique for American breweries.[37]

Brewery architecture became a specialized profession and few general architects would attempt to design a brewery any longer. Among the noteworthy brewery architectural firms were August Maritzen; Widmann, Walsh and Boisselier of St. Louis; and Frederick Wolf and Louis Lehle of Chicago.[38] German-born brewers sought out German-born architects who would understand their needs and wishes.[39]

The Fred W. Wolf Co. of Chicago began in 1867, continued until Wolf's death in 1912, and then the derivative Wolf-

Figure 2.6. Crescent Brewery of Nampa, Idaho, built in 1906 in the Romanesque style. (Photo from author's collection.)

Linde Company continued to 1998. The company's massive archive of drawings is now at the Chicago Historical Society.[40] The company specialized in refrigeration equipment and often designed such installations in existing breweries or in new breweries that had been designed by other architects. Work by the company was performed across nearly the entire United States; the firm's documents would be invaluable to anyone researching or excavating a brewery designed by this company in whole or in part.

The architecture of Chicago breweries was unique, being built in the home location of the best brewery architects in America, and therefore adjacent to the best thinking and developments in the field. Chicago breweries of each era would be expected to be perfect examples of each style that in turn dominated the field. The *Rundbogenstil* style was obvious in the Schoenhofen Brewery on South Canal as well as the Carl Corper Brewery on South Union that remained extant into the twenty-first century.[41] Hugh M. G. Garden designed the Schoenhofen warehouse and powerhouse, however, in Prairie style in 1902.[42] Other Chicago breweries were designed in Italianate and Greco-Roman styles from 1870 to about 1900. After 1900 the ornate style lost favor in Chicago. The Crescent Brewery of Nampa, Idaho, built in 1906, was still in the Romanesque style (fig. 2.6).

Brewery Architecture

Brewery architecture historian Susan Appel divided the brewery architects into a first generation of brewery architects practicing from 1865 to 1885, a second generation of pre-Prohibition architects from 1885 to 1900, and a third generation from 1895 to 1920. After Prohibition this architectural specialty never returned.

Cellerage and Above-Ground Icehouses

Until about 1850, beer brewing had to be coordinated with the seasons of the year to control brewing temperatures. Ale was not brewed during the hottest months but otherwise, ale brewers did not much consider temperature. Lager was brewed only in the cold months and stored—*lager* means storage—until the hot months when it was in greatest demand. Storage had to be done in cool to cold—never freezing—areas to keep the lager in good condition.

Beer Cellars

Underground storage afforded natural insulation. Brewers would place ten- to fifteen-barrel casks on skids in caves. The caves were carved with arched ceilings and divided into sections so the whole cave would not get warm while one section was being emptied.

When these proved inadequately cool for lager beer, ice chambers were added above the caves. The great weight of the ice required extremely strong structures. Merely storing large quantities of ice proved insufficient, so gradually, aboveground cellars that factored in ventilation, humidity, and other environmental issues developed. Some of these cellars used ice with salt to lower the temperature and required small amounts of ice to be added daily to replace that melted by the salt.

A detailed description of small-scale cellar making exists. Henry Mayer of Collinsville, Illinois, confided to his diary in 1858 his efforts in making and keeping in operation his beer cellar. "I have been to Belleville [Illinois] and seen how to build a beer cellar; in a short time I will commence to build a cellar in my field near the saw mill, 100 feet long, 17 feet wide and 14 feet high; I calculate it to cost 1000$ when finished complete."[43] Mayer's diary entries chronicled the progress of the project.

> **27 August 1858:** Three men were working on the cellar.
>
> **[A month later:]** They are still digging the cellar, have gone in 40 feet on each side . . . will begin on Tuesday to lay brick for the side walls.
>
> **29 September 1858:** Mr. Withers and the two Dutchmen have given up the cellar and I have to get others; they wished to force me into measures that I did not like and that they, at first did not ask.
>
> **17 October 1858:** To day about noon they finished the first arch of 16 feet and are now removing the dirt when the ribs, braces, etc. will be taken out and you can see the size of the cellar.
>
> **23 October 1858:** Yesterday the men finished digging 16 feet more, set the arches and Mr. Kuhn worked half of today laying brick. If Mr. Withers, Charley Flanders and old Kuhn work next week they should finish by Wednesday; Thursday and Friday they should remove the dirt with a cart and by Monday following (1st Nov) be digging out the next 32 feet.
>
> **16 November 1858:** Have been busy all day hauling brick from Kneedler's; Mr. Hogg will haul tomorrow at 75 cents per 1000 and as soon as I am done threshing will haul the balance and sand.
>
> **22 November 1858:** Last week, the bricklayers built the walls running back the distance of 64 feet & also dug out the ground for 16 feet & put up the ribs, today they finished nearly half the arch & if I can get about a barrel of lime will finish it tomorrow, they will then haul out the dirt in 3 days & I think cut the balance of the dirt for the remaining 16 feet; there then will be completed 64 feet and I intend to have 16 feet more making 80 feet in a straight line; at which point I shall stop till next summer, or till I see how the brewery business turns out.[44]

Mayer's belief that his work on the cellar was concluded would prove to be false. His diary would later reveal complications with the cellar work. He continued to make many notes, including a record of the equipment he installed. Eventually, his new cellar would demand more attention than he expected.

> **6 January 1859:** . . . a Pump has been fixed stationary in the room over the first cellar & we now pump the ale from a Small Tub in that cellar into a big cask that stands upon a wagon outside the window . . .
>
> **17 April 1859:** Last Wednesday . . . a rain fell about 11 o'clock which was harder than was ever known at this place before; it overflowed the bank of the Cave and ran into the cellar about a foot deep; the two outside walls of the shed fell in; I have been busy since Thursday morning having it repaired will finish tomorrow by noon; it will cost me about 30$ to make it as before; but will be about 50$ better for I am having the Walls built differently and to exclude all the air.
>
> **19 January 1860:** On Monday night (16th) the back wall of the cellar fell in with a large quantity of dirt. I have had 4 men at work clearing it away.[45]

Clearly, cellar construction was not for amateurs, and not without pitfalls.

Icehouses

American icehouses could have ice twenty feet high on top and exert a weight of 1,150 pounds per square foot on the floor or lower surface.[46] As much space in the beer storage area was devoted to ice as to beer. Various architects and engineers began to develop specialized icehouses. In 1882, David W. Davis installed his patented "Non Plus Ultra Ice-House" in Detroit at the Voight brewery.[47] Using salt on the ice, the temperature was kept very low, less ice was used, and minimal labor was required.

Frederick Colyer recommended icehouse construction of timber boarding on the inside and outside of framing, and "filling the interstice with sawdust, thatching the roof, care being taken to provide good drainage, as damp is as much to be feared as heat."[48]

Joseph Dudler of the Salt Lake City area built a cellar sometime before 1875 into the hillside near his inn and brewery; it was 15 feet wide, 20 feet long, and 10 feet high.[49] The entrance measured 6 feet by 4 feet. The structure was still extant as of this writing. In 1887, the Idaho Falls Brewery built a cellar measuring 28 by 50 by 12 feet that could hold 700 tons of ice.[50] The report gave no particulars of the construction techniques. In 1893, William Von Berge of Grangeville, Idaho, built an icehouse at his brewery with brick walls 18 inches thick and a brick drain.[51]

Makeshift arrangements for storing ice were possible. For instance, John Lemp, a Boise, Idaho, brewer, emptied his icehouse completely by mid-September, 1891. He was then getting ice out of a brick elevator nearby where excess ice harvest from the previous winter had been successfully stored.[52]

Refrigeration ended the need to put tons of ice on top of a cellar and thereby changed brewery architecture by allowing more numerous, less bulky above-ground cellars, also called stock houses. These usually lacked windows to aid in insulation, which gave them a more solid appearance than an adjacent brew house with windows.

The underground beer vaults or cellars of the Phoenix Brewery of Louisville, Kentucky, were 100 feet long and 25 feet wide, with arched ceilings 15 feet high. They were abandoned and the entrances grown over but still visible in 1995.[53] Many such structures must await investigation around the United States.

Examples of Brewery Construction

The Carson City Brewery in Nevada, built about 1860, was native limestone and measured 20 by 40 feet with walls 20 inches thick. "Legend has it that between the ceiling and the roof it was filled with sod for insulation. A horse barn was built on the corner from rock and cement. It had sliding doors on the north and south ends of the building, and a hayloft above."[54]

Ernst Weisgerber left his brewery job in Vancouver, Washington, to locate at Lewiston, Idaho, in the spring of 1862. As lumber was scarce in Lewiston, he built a rough brewery of boards running up and down—an architectural disaster but upper middle class in a city of tents.[55]

In 1864, the malt house of the Kessler Brewery in Helena, Montana, was built of hand-hewn native rock. Eighty-five years later, in the 1950s, that building was coated with stucco and served as the cooperage-pitching department.[56]

The *Tri-Weekly Statesman* of Boise, Idaho, glowed with praise over a cellar that Professor James Hurd had just built for the John Lemp brewery in 1868. It measured 16 by 42 feet with an arch of "11 feet in the clear" of native stone from the bluffs east of town and was "superior to anything of its kind in the territory." [57]

In February, 1891, disaster struck the Koeninger brewery of Ketchum, Idaho.[58] A fire destroyed the three-story log brewery, fifty to sixty cords of fuel wood, and a number of outbuildings. No photos are known to exist of this structure, but a three-story log manufacturing plant of any kind would be a curiosity in the 1890s. Few breweries were log structures.

The Klug Brewery building at Challis, Idaho, has lasted since 1879 and is on the National Register of Historical Places. The credit for the brewery's longevity goes to its building material—stone. The builder is also awarded some of the credit; professional stonemason George Fuchs, co-owner of the brewery with Fred Albiez and Ferdinand Klug, was an excellent choice to oversee construction. Fuchs was the junior brewing partner but he had learned in his native Germany how to build the structure they needed for brewing.[59] The building was only a brewery for a few years but it had numerous other uses over the next decades. Archaeological excavations were conducted around the building in the 1980s.

In 1893, William Von Berge wanted to build a wonderful new brick brewery for Grangeville, Idaho.[60] This brick brewery was a large project requiring time and capital for completion. Von Berge started excavations in January, and bought the two-acre orchard across the street for working space.[61] This was to be the first brick building in the county. The projected brewery would be able to produce 150 barrels per day.[62] Excavations continued, and by August, 1893, the foundation was being laid, but Von Berge's attention had been diverted. While digging, a bed of sand was found that was suitable for brick and mortar making.[63] Von Berge started a brickyard on the former orchard property. Sand came from across the street, clay from the orchard itself, and the mill ditch brought the water. The newspaper said that good brick had long been needed in Grangeville. Von Berge faced an opportunity for a second business.

The formula for the bricks included lime. According to a newspaper account, a Mr. Norton left for his limekiln on the south fork of the Clearwater River to burn lime for Von Berge's bricks.[64] A week later, Von Berge was ready to fire a kiln of 50,000 bricks.[65] In October, the bricks were finished and the ones not needed for the brewery plan were offered for sale to the public. They were the "toughest brick ever made on the prairie [Camas Prairie around Grangeville]."[66] In December, the cellar and icehouse with its 18-inch thick walls were completed and work was progressing on a brick drain out of those rooms. The paper said it would be "a very permanent structure and ornament to the town when completed next spring."[67]

A number of towns in Idaho had not a single brick building until a brewery was built of that material. Once the practicality of brick was demonstrated, many other brick buildings went up.

During 1909, Fritz Schleifer, a brewer in Silver City, Idaho, announced to the brewing world that he had put cement floors and cement walls in his malting room.[68] Seventy-eight years later, while in Silver City on an unrelated archaeological excavation, I discovered that the brewery building was long gone but those cements walls and floor were still there and easy to locate.

The Idaho Falls Brewing Company underwent major construction in 1908. The new fireproof brewery had a capacity of 25,000 barrels per year. The plan was compact yet commodious, with every wall used twice. The foundation was concrete with brick walls and cut stone trimming, and the floors were asphalt on concrete arches between steel beams. Bernard Barthel, a Chicago brewery architect, designed the plant. Major companies supplied equipment and materials: Goesty and Floodin Manufacturing Company of Chicago provided the copper and tank work; Kaestner and Company of Chicago handled machinery and millwright work; Thaler Brothers of Chicago made the chip casks; Bilter Manufacturing Company of Milwaukee supplied the refrigeration machines; Armstrong Cork Company of Pittsburgh sold them cork insulation; and Schillinger Brothers of Chicago supplied asphalt and roofing material.[69]

In contrast to medium-size brewing plants, large breweries were housed in giant complexes that were marvels to behold. Here is a description of the Chris Stahlman Brewery of St. Paul as of 1882:

> The Extensive establishment of the Christ. Stahlman Brewing Company is situated on the corner of Fort and Onedia Streets and commenced business July 5, 1855. It consists of five three story buildings, built of stone, and two basements, covering an area of sixty odd lots, the total cost of lots and buildings amounting to $150,000, beneath which are cellars and compartments three stories deep, covering an area of one mile in extent, and extending beneath all these buildings and under the street, over which the street cars pass daily. The works under ground cost $50,000, and are more wonderful in their aptitude and construction than the buildings above ground, and where most of the work is done. Beneath the ground in these various compartments, are to be found the malting, drying, fermenting, and storage rooms. There are eight malt rooms, two drying rooms, fifty tubs for fermenting and 200 hogsheads for storing beer, containing about 1,000 gallons each. There are also two artesian wells connected with the brewery, one 315 and the other 217 feet below the surface of the ground, with a capacity of 2,600 barrels of water per hour. Attached to these are two steam pumps and iron pipes of three and five inches in diameter, for conveying water and beer throughout the entire establishment, and also have fireplug connections. There are two large boiling tanks, making over 110 barrels of beer per day. The cost of coal fuel per month is $500. For hops and grain $200,000. The machinery of this vast concern is driven by two steam engines of 25-horsepower each, and the three boilers 35 to 48-horsepower. Their average amount of business is over one-quarter of a million dollars. They employ forty-seven hands in their business, at a monthly cost of $2,000. Besides their city trade, which is extensive in itself, their business ramifications extend throughout Minnesota, Iowa, Wisconsin and Dakota Territory.[70]

The malt kiln of the Lemp Brewery of St. Louis, in late 1877, had three very large furnaces, and the kiln structure was separated from the main building by a thick partition wall built of brick, iron, and fire-clay tiles and built to be as fireproof as humanly possible.[71]

In 1912, Anheuser-Busch proudly advertised its size. The St. Louis brewery covered 140 acres, equal to seventy city blocks, with 110 individual buildings. It had a malting plant, grain storage, stockhouses, steam plant, electric plant, refrigerator plant, and ice plant.[72] This was the zenith of brewery material culture.

Beer Gardens

The beer garden, derived from the German *Bier Garten*, was a feature often associated with lager breweries. A German tradition created space to have an outdoor area for families to eat, drink, and sometimes listen to bands.

An 1850 lithograph of a beer garden included a notice that patrons were limited to forty-eight glasses of beer.[73] Perhaps this was just a joke, but perhaps not. By the 1850s German beer gardens were common in St. Louis. The Klausmann Brewery there added one in the 1860s, and Joseph Schnaider at his Choteau Avenue Brewery built one in 1865.[74] Some institutions called beer gardens sold beer but were not associated with breweries, and some places called beer gardens were actually indoor halls with no gardenlike aspect at all.

In Cleveland, some beer gardens had nothing but indoor facilities by 1857.[75] In the early 1850s, John Dangeleisen started one of the first true gardens there. A favorite pastime at Dangeleisen's was "Kind und Kegel" (skittles).

Frederick Halnorth established the most popular beer garden in that city in 1862. At its peak of business, it had an enormous concert hall, two restaurants, a museum, bowling alley, rifle range, photo gallery, and a fishing pond spanned with a picturesque bridge. By the 1870s real beer gardens outside the city limits avoided the Sunday closure laws of

Cleveland proper. The 1875 Andreas Atlas of Iowa showed only one beer garden in the state: The Sioux City Brewery.[76]

The popularity of beer gardens was directly linked to the public change in preference from ales to lagers. In 1880, F. W. Salem revealed his opinion of why some people still preferred ale over lager: "others because the comparatively secluded and unsocial character imported from England to our ale-houses suits them better than the more social and gregarious customs of the lager beer garden."[77]

A Cincinnati description from 1883 elaborated on the reasons that lager had grown so popular:

> . . .in no little degree, at the same time, to the taste which has been acquired for "Lager," as a beverage, not only among the native German population, but all classes. Beer Gardens, where this beverage is swallowed by old and young and in incredible quantities, have become institutions of great magnitude in this and all the large cities of the Union, and so necessary has the article become in the estimation of the masses, as a beverage, that a scarcity of "Lager" is regarded in a similar light with a dearth of corn, and from the fondness displayed for it, and the rush made to those shops, when a fresh keg of 'Lager' has been tapped, the conclusion seems natural that a large number of citizens would dispense with their bread rather than their beer.[78]

Virginia City, Nevada, had a German-style beer garden started by an American-born man named van Bokkelin.[79] Goldfield, Nevada, one of the largest and fastest growing gold camps in the history of the West, had the Goldfield Brewing Co. from 1905 to 1908. The Goldfield brewery produced steam beer that reportedly tasted rather peculiar. It had a European beer garden where "working" girls used to come early in the evening. In March, 1912, this brewery was owned by the Reno Brewing Company but was not in operation.[80]

The tradition of the beer garden is alive and well, even in recent years. The Capital Brewery in Madison, Wisconsin, had a beer garden next door as of 1992.[81]

Breweries on Fire Insurance Maps

Thousands of depictions of breweries can be found on the fire insurance maps drawn for American cities and towns. In 1788, insurance mapmakers created a map of Charleston, South Carolina.[82] By the middle of the nineteenth century, such maps were common as insurance companies and underwriters sought to calculate risks, and lithography printing made reproduction of the maps reasonably easy.

After World War II the use of the maps declined, and by 1960, the National Bureau of Fire Underwriter's Map Committee noted that most companies no longer used such maps. During the century or so in which these maps were in vogue, they were and remain an invaluable source for the physical description of breweries.

A number of cartographic publishers existed in the early years, but after 1876, when the Sanborn Map Company was incorporated, the business consolidated. Sanborn maps for some 12,000 cities and towns in every state, and a few Canadian and Mexican cities, were drawn. The Library of Congress holds over 750,000 sheets of Sanborn maps, and many regional libraries hold copies.[83]

These maps depict the building type, using different colors for frame, brick, stone, iron, or adobe construction. The maps show the location of brew kettles, grain hoppers, icehouses, and so on, and are all drawn to scale. Maps often list the names of owners, the use or nonuse of electricity, the type of fuel used for heating kettles, the length of fire hoses, and the number of people who sleep in the building. Often, no other written sources show such things. One of the most valuable ways to use the maps for historical purposes is to compare the brewery structures on different maps to see the changes over time.

I found an excellent example of this use in my research on the Lewiston, Idaho, brewery. Maps for 1888, 1891, 1896, 1900, 1904, 1909, and 1918 all depicted the location and its structures. The 1888 brewery had a straight-sided, rectangular building about six times as long as it is wide. A saloon stood in front. Also visible were a beer cellar, a water tank on the roof, a one-and-a-half story room with a fermenting cellar, and a malt cellar. A third space housed the kettle enclosed in one corner, a mash tub, and a furnace. The malt kiln was apparently a separate structure that abutted the brewery. The complex had a separate icehouse and a separate pump house with a windmill and a malt mill (figs. 2.7, 2.8, 2.9, 2.10, 2.11, 2.12, 2.13).

The 1891 map showed a boiler addition on the pump house with a vertical steam boiler and a 10-foot iron chimney but no other changes. The 1896 map showed an addition that ran the entire length of the brewery so that the malt kiln was then enclosed and an attached one-story icehouse on a stone basement stood at the front. At this point, a vertical steam boiler in a one-story L-shaped addition stood adjacent to the brewing and mashing room to the south. The pump house and windmill were gone entirely.

The 1900 map showed new buildings. The main brewery was three stories high and roughly square. The northern half was designated for fermenting on the second floor, with beer cellars on the first floor and in the basement. The south half of the building had the brew kettle on the first floor, mash tubs on the second, and the malt mill and grain storage on the third. Probably the entire third floor was used in this way, since no other use for the space was noted. There was also a water tank on the roof at the southeast corner of the building. The general arrangement indicated quite clearly that the gravity-flow brewing process was being used. The two-story office replaced the saloon, and the artificial ice-manufacturing machine was shown along with other changes.

The 1904 map showed only the addition of a railroad spur to the rear of the building. The 1909 map revealed great developments with a new bottling department, new wash-house for bottles, and an additional grain storage area. The 1918 maps clearly illustrated that Prohibition had struck. The equipment had been moved out and the building was partially used as a grocery warehouse. A great deal of more specific information could be coaxed from these images. Maps such as these constitute a necessary pre-excavation research tool, if they exist, for a given brewery site.

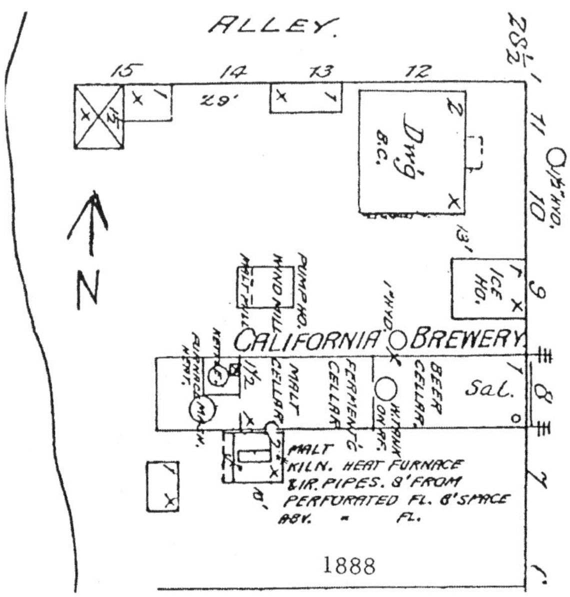

Figure 2.7. 1888 Sanborn Fire Insurance map of Lewiston, Idaho, that shows the brewery location. (Sanborn maps.)

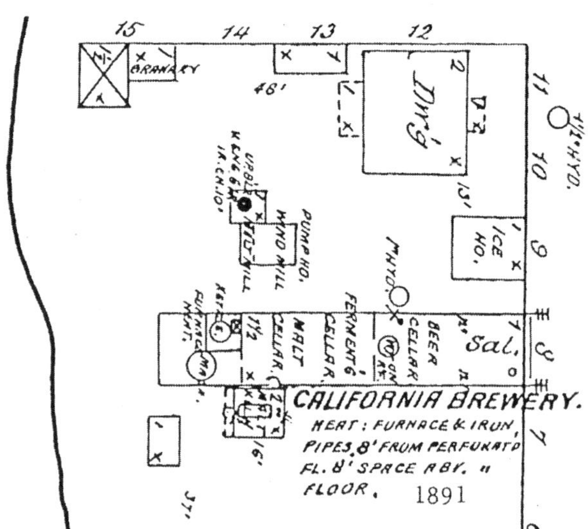

Figure 2.8. 1891 Sanborn Fire Insurance map of Lewiston, Idaho, that shows the brewery location. (Sanborn maps.)

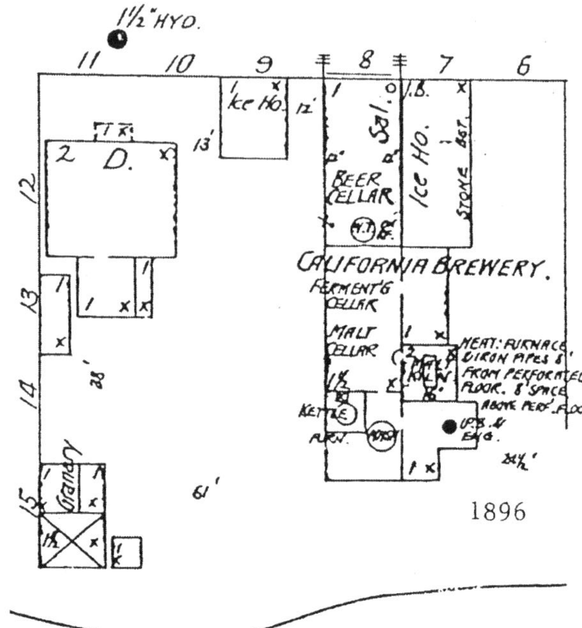

Figure 2.9. 1896 Sanborn Fire Insurance map of Lewiston, Idaho, that shows the brewery location. (Sanborn maps.)

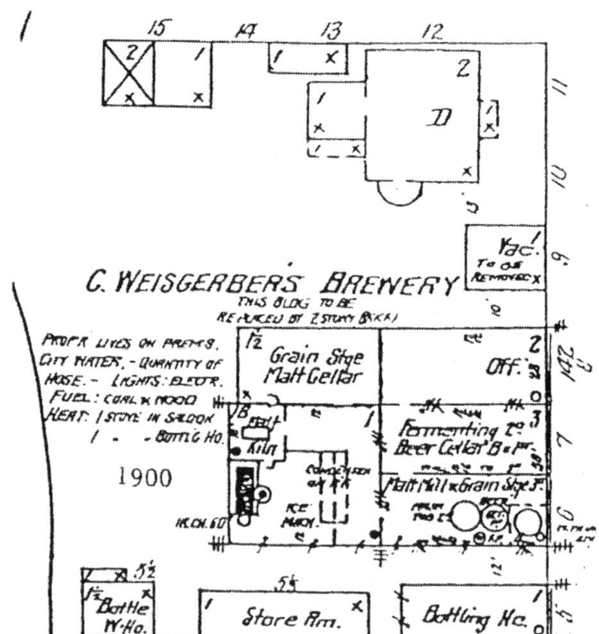

Figure 2.10. 1900 Sanborn Fire Insurance map of Lewiston, Idaho, that shows the brewery location. (Sanborn maps.)

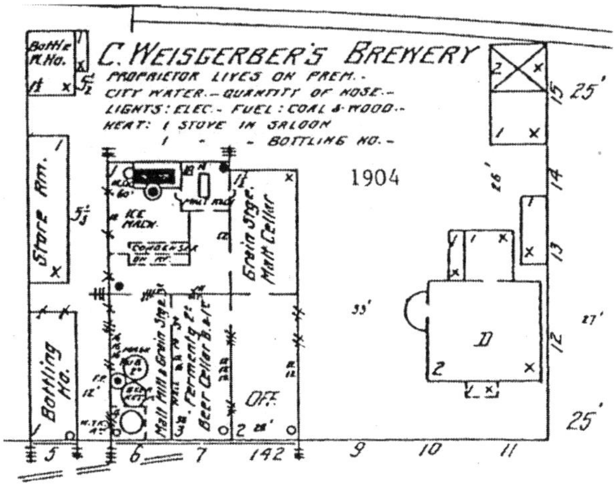

Figure 2.11. 1904 Sanborn Fire Insurance map of Lewiston, Idaho, that shows the brewery location. (Sanborn maps.)

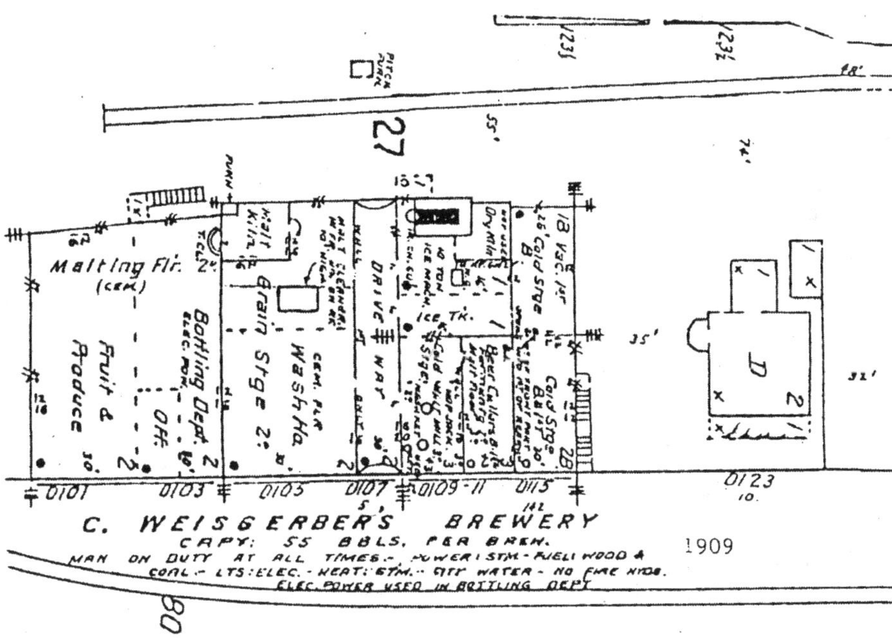

Figure 2.12. 1909 Sanborn Fire Insurance map of Lewiston, Idaho, that shows the brewery location. (Sanborn maps.)

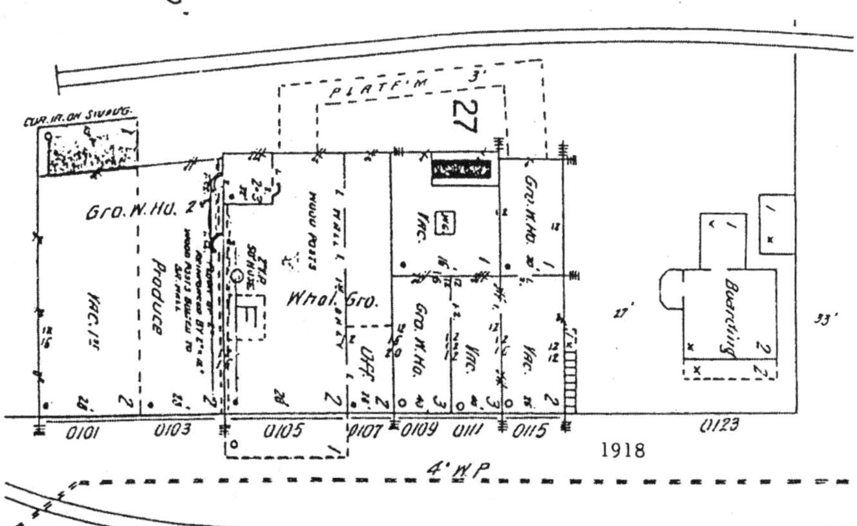

Figure 2.13. 1918 Sanborn Fire Insurance map of Lewiston, Idaho, that shows the brewery location. (Sanborn maps.)

Fire, Earthquake, and Tornado: Destruction of Breweries

Breweries were at high risk for fire due to the boiler for the steam engine, the malt kiln, and the explosive potential of the malt mill. More brewery fires occurred than could be recorded. Brewers sought protection and were well represented on fire departments in the West. In fact, a brewery was often the principal impetus behind the formation of a fire department in its town.[84]

In 1881, John Hauenstein's brewery in New Ulm, Minnesota, was destroyed by a tornado. The beer in the cellar was not damaged, so sales continued. Hauenstein brewed under the open sky while the replacement brewery was under construction.[85]

In 1935, earthquakes in Helena, Montana, nearly destroyed the Kessler Brewery. The Northern Pacific Railroad shunted a locomotive onto the Kessler siding, and steam from its boilers was piped into the brewery to keep it operating in the zero degree weather. The brewery stayed in operation and made repairs costing $62,500.[86]

On August 7, 1909, the old and unused John Lemp Brewery of Boise, Idaho, burned.[87] The oldest part of the building, dating back forty-five years, was completely destroyed. The local newspaper reported that Officer Bostwick noticed the fire a few minutes before 3 o'clock in the morning and gave the alarm.[88] Hundreds of people gathered to watch the spectacular blaze.

The fire department arrived quickly but found the sheds and storage building completely in flames. They had no chance to save anything on the Main Street side. They directed their efforts to saving the granary and the stables, which they succeeded in doing. The big brick icehouse shielded the Lincoln School, which was adjacent. The horses in the corral were freed in time. A shed at the nearby Ridenbaugh's lumber yard caught fire but the flames were quickly extinguished. "The prevailing opinion was to the effect that the old buildings were an eye sore to the community and there was little regret that they were destroyed."[89]

The John Natter Brewery in Heppner, Oregon, was destroyed in a famous flood that took much of the town in 1903. Idaho City, Idaho, suffered three major fires in the 1860s and breweries were destroyed in each. A psychotic arsonist burned the Moscow Brewery in Idaho in the summer of 1908. Jacob Weinhard of Dayton, Washington, nephew of Portland, Oregon, beer baron Henry Weinhard, was burned out four times in his first seven years in Dayton. This may be a record, yet he stayed and prospered. The brewery of Harpers Ferry, West Virginia, which has been extensively studied, burned three times, had its smokestack knocked down by a cyclone, and was flooded by the adjacent river several times.[90] A list like this could go on and on. Many breweries were reduced to ruins that still lie beneath the ground.

Idaho Brewers Associated with Fire Departments

Richard Adelmann	Boise, Idaho
John Weisgerber	Lewiston, Idaho
Carl Mallon	Wallace, Idaho
John Brodbeck	Boise, Idaho
Godfrey Gamble	Yreka, California (before moving to Idaho)
George Gans	Idaho City, Idaho
Joe Helmuth	Walla Walla, Washington (before Idaho residency)
George Kohlepp	Boise and Hailey, Idaho
John Lemp	Boise, Idaho
Jacob Lockman	Wallace, Idaho
Joseph Misseldt	Boise, Idaho
Henry Reiniger	Rathdrum, Idaho

Bottling Houses

Bottling went from a small activity carried on by one worker in a small shed near the brewery to a giant activity requiring specialized buildings with great capacity. In about 1872 to 1873, Anheuser-Busch began to bottle enormous amounts of pasteurized beer and needed enlarged facilities. A new two-story building, measuring 40 by 180 feet and housing three bottling machines, turned out 40,000 bottles a day. A second two-story building, 109 feet by 112 feet, went up almost immediately and increased capacity to 100,000 bottles per day.[91]

Bottling houses were generally long, low, rectangular buildings with one main floor filled with bottling machines. These were separate buildings from the brewery because of the tax law in effect up to 1890 that stated that beer had to be kegged and the tax stamps affixed before being transported over a public road to be bottled.

When the law changed, bottling houses did not change noticeably. The bottling house at the Moerlein Brewery, Cincinnati, built in 1912, was 225 feet by 117 feet, with a government cellar (the place where the beer was metered for tax purposes) in the basement, a main bottling floor, and an upper floor for storage with skylights.[92] It was made of reinforced concrete faced with brick.

Tiny breweries also bottled their beer. For instance, in 1909, Charlie Brown and his son Everett built the bottling building for the Palouse, Washington brewery.[93] It had a front door and two front windows, with the walls left windowless for storing barrels. The six men employed in the building used Palouse River water to wash the floors. The brewery capacity was one hundred barrels per brew and the facility usually brewed three times per week. Beer not sold locally was shipped by rail into small Idaho towns of Bovill, Fernwood, Clarkia, and Elk River. The brick bottling plant was extant in 2007.

The greatest pre-Prohibition bottling building was again at Anheuser-Busch in St. Louis. The Bevo bottling plant, named after the company's near beer Bevo, was built between 1916 and 1918. It still stands, is in use, and is proudly shown to the thousands who tour the brewery every summer. It measures 580 feet by 250 feet and stands eight stories high, with 1.2 million square feet of floor space. Thirteen railroad tracks run through the basement that can hold ten cars each. At maximum capacity, two million bottles per day, or 125 carloads, could be shipped.[94]

Structures Built in Association with Breweries

In addition to the buildings needed for the basic production of beer, many breweries erected structures for specialized purposes that became part of the overall business conducted at the site—sometimes, purposes with no inherent connection to brewing. The custom at one time was for the owner to erect an impressive dwelling very close to the brewery. Brewers' mansions often ended up surrounded by industrial buildings.

A wide variety of structures and features have been located on brewery property. Known features of breweries have included:

- Advertising signs
- Architectural decorations such as statues or giant clocks
- Beer gardens
- Blacksmith shops, including horseshoeing facilities
- Bottling works
- Bowling alleys
- Carpenter shops
- Cattle and hog pens, corrals, and so on
- Caves
- Coal yards or wooden storage sheds
- Cooperage shops
- Icehouses
- Ice ponds
- Laboratories
- Machine shops
- Malting facilities
- Pitch houses
- Railroad sidings or spurs
- Saloons
- Springhouses
- Stables, including storage of horse feed and gear
- Steam engines
- Storage facilities, including icehouses and stables, near railroads
- Truck garages, including paint shop, for gasoline and electric trucks
- Wells
- Windmills

A number of examples are known of related and unrelated brewery property structures. At Grangeville, Idaho, in 1880, brewer William Von Berge built a 120-foot by 150-foot ice pond and stocked it with carp, so he could cut ice off the pond in the winter and catch fish in summer.[95] The pond and the icehouse are both brewery-related structures. In the vicinity of Washoe City, Nevada, the Chauncey Haskell Ice Pond was studied in 1998 by the Kautz Environmental Consultants.[96]

Figure 2.14. The lighted Bohemian Club sign on the roof of the brewery in its closing days of 1961. (Photo by author.)

The artificial pond was 12,840 square meters, situated on a hillside northwest of town. The ice was much in demand in the 1870s.

In 1882, the Weisgerber brothers of Lewiston, Idaho, added a windmill—"a patent self reefer"—to their brewery.[97] Due to their location at the confluence of the Snake and Clearwater Rivers, their well must not have been very deep, but this was a free source of power. By 1896 the windmill was gone. In October, 1905, the North Pacific railroad spur at the Weisgerber Brewery was extended to reach an additional building that had just been completed.[98]

George and Jacob Gans opened the Bannock Bowling Alleys at their brewery in the tiny mining camp of Buena Vista Bar, Idaho, in 1863.[99] In 1892, William Von Berge, the Grangeville brewer, built a 64-foot-long bowling alley attached to his brewery.[100]

A large electrical sign was erected on the roof of Montana's Billings Brewing Company in 1933. It ingeniously displayed a bottle of its Old Fashion [*sic*] beer pouring its contents into a mug. Unfortunately, a steam pipe on the roof sometimes produced condensation and caused the wires to short and extinguish the 2,000 bulbs the sign employed.[101] Many breweries had large, sometimes lighted, signs on the roof or near the brewery (fig. 2.14).

Caves Used by Breweries

Caves were used for storing beer in Europe and in several places in the United States. The term "cave" has a number of connotations in popular usage, and may be interchangeable with the terms "vault," "cellar," and "tunnel," or even "crevice."[102] Ingalls has noted that "cave" in eastern Iowa may refer to a natural or man-made underground opening. Many caves were constructed or modified from their natural condition for use as root cellars, ice storage, wine storage, or cisterns. These may not differ substantially from brewery caves. Ingalls also noted about twelve other caves in the same area as the Law Cave that he studied.[103]

The Chris Stahlman brewery of St. Paul, Minnesota, was known locally as the cave brewery because it used caves along the Mississippi River.[104] The man-made cave at the Miller brewery in Milwaukee was used as a bomb shelter in the 1950s, and is still part of the brewery tour.[105]

St. Louis, Missouri, though, deserves credit for the most significant brewing caves in the United States. Adam Lemp, the first lager brewer in St. Louis and founder of a huge brewery that eventually shipped beer throughout the nation, discovered a cave on the road to Carondelet and built his second and much enlarged brewery above it.[106] He flattened the ancient river courses and put in masonry walls to divide the cave into compartments. Dr. Nicholas DeMenil built a house nearby, also over the cave, and used his portion for storage and leased space in it to the rival Minnehaha Brewery.

Lemp then built his mansion near DeMenil's home. Lemp ran three pipelines—cold water, hot water, and beer—from the brewery, through the cave, and to his home. A nineteenth century visitor to Lemp's cave described it as 700 feet long, with eight hundred to one thousand barrels of beer stored in it. These barrels were probably large tuns or hogsheads that held many barrels each.

The two Uhrig brothers who owned the Uhrig brewery bought land above a cave in St. Louis in 1852. They connected several smaller caves and installed brick walls, an arched ceiling, and a narrow gauge rail to transport the beer. They also saw retail potential and put in picnic tables, held band concerts and picnics, and offered tours of the cave.

Julius Winkelmeyer's St. Louis Brewery was the largest producer in the city by 1850.[107] Beginning in 1849, he dug two underground cellars, one below the other, using a natural cave. These were 90 feet long and parallel to Market Street. Later descriptions, after some remodeling, noted four-foot-thick masonry walls, stone and brick columns, and in the lower chamber, which sat on bedrock, brick and masonry vaulted arches. In 1933, equipment was still in place indicating the upper level was the fermentation cellar and the bottom had been for lagering the beer.[108]

The most famous St. Louis brewery, Anheuser-Busch, owed the location of its brewery to the presence of a cave. George Schneider built a brewery over the caverns in 1852. He went broke and Urban & Hammer took over and expanded, but they were taken over by Eberhard Anheuser when they became insolvent. During the Civil War, guns from the federal arsenal were hidden in the caves. The brewery became Anheuser-Busch, artificial refrigeration arrived, and the caves were forgotten. In the 1930s, the caves were rediscovered. At least eleven other St. Louis breweries used various caves in the city. Only the roughest details about these features have survived.

Livestock Raising at Breweries

Many brewers raised pigeons, hogs, and/or cattle at the brewery. Michael Diversey, an Alsatian who arrived in Chicago in 1832, operated a dairy out of the Haas brewery.[109] The two operations shared stored ice and their best business seasons varied, which tended to balance out the income through the year. The animals could be fed spent grain easily and inexpensively. Theodore Hamm in St. Paul had sheds for his cattle and pens for his pigs. These supplied meat to the Hamm family, the brewery employees, and the beer garden.[110] He had as many as forty steers in fattening stalls.[111]

John Lemp of Boise sold $600 worth of hogs in June, 1877. The hogs had been fed the spent grain from the brewery, making them extremely profitable. Lemp vowed to get rid of all his hogs and pigeons because the town had grown up around the brewery and the smell was annoying to the new neighbors.[112] Years passed before he actually accomplished this.

Brewery Warehouses Along Railroads

As railroads proliferated, brewers began to set up branch offices with beer storage facilities over ever-widening sales territories. A depot would receive beer in kegs and barrels and then deliver it to saloons. A depot generally had an icehouse, warehouse, office, and stable for the delivery horses.[113]

Beer retailers generally had limited storage space and ordered about a week's supply of beer at a time. "Although the total beer order of several retailers in a particular area might fill a carload, Adolphus Busch found the costs prohibitive to ship partially filled cars."[114] The solution was to build storage facilities near railroads to hold the retailers' extra beer until ordered.

After completion of the Chicago and Northwestern Railway linking that city with Milwaukee in 1855, Best (Pabst), followed by Schlitz and Blatz, all established warehouses for their beer in the Windy City.[115]

In Kansas City, depots were soon maintained by Anheuser-Busch, Val Blatz, Philip Best (later Pabst), Dick Brothers, Wm J. Lemp, and Joseph Schlitz.[116] "All in all, some twenty-two breweries from six states tried to penetrate the Kansas City beer market in the years prior to Prohibition."[117]

Professor Elliot West noted that, during the last two decades of the nineteenth century, agents for large breweries in California, Missouri, and Wisconsin fanned out through the Rocky Mountain states marketing their beer. For instance, "after the arrival of the railroad in Helena [Montana], one distributor sold 1,600 kegs of Schlitz alone between April and July."[118] In 1879, the Leadville, Colorado, agent for Anheuser-Busch and Company had "contracted with the railroad to keep at his disposal his own crew and train of seven cars so his warehouse would never be empty."

Out of sheer necessity, the breweries built cold storage warehouses near the railroad to store the carloads of beer. For instance, Schlitz erected a warehouse at Hailey, Idaho, in 1883, on the heels of the arrival of the Oregon Short Line Railroad.[119] Schlitz built another at Caldwell, Idaho, in 1891. Gund Brewing of La Crosse, Wisconsin, built a warehouse at Idaho Falls in 1901, and by 1907, five various breweries had storage facilities in Lewiston, Idaho. Salt Lake Brewing Company put a warehouse in Hailey in 1906 that could hold two car loads of beer. The Fred Fug brewery of Omaha, Olympia of Tumwater, Washington, and others were also shipping into Idaho.

An indication of the size and relevance of these buildings was apparent in 1891. That year, a refrigerator belonging to Pabst Brewing in Idaho Falls had a property tax assessment of $250, one belonging to Anheuser-Busch was assessed $200, and Schlitz's new facility was assessed $500.[120] Also in 1891, Louis Elg, agent for Anthony and Kuhn of the St. Louis Brewing Association, was building a large icehouse in Idaho Falls. He was "portioning his ice house, placing in a solid floor, sides and ceiling, which will be completely surrounded with ice."[121] The Seattle Brewing and Malting Company built a 20-foot by 40-foot icehouse along the railroad in Wallace, Idaho, in 1907, just after Inland Brewing of Spokane had built a 24-foot by 42-foot building nearby.[122]

Business Conducted in Breweries During the Prohibition Era

When Prohibition went into law in 1919, the brewery buildings and their workforces, as well as the investors and management personnel, were all in need of alternative ways to make money. Evidence of the building and equipment modifications of this era may still be part of the recoverable physical record. No comprehensive list of the Prohibition-era businesses is possible, but the material that follows gives some idea of what might be encountered. Also, breweries conducted sideline businesses during the best of times, and this description gives an idea of what might have been found during other time periods.

Near beer—beer with one-half of 1 percent alcohol—was an obvious replacement for beer, and many brewers had begun experimenting with it before being forced to abandon beer making by Prohibition laws. Among the near beer brands were Bevo, Yip, Ona, Pablo, Hoppy, Famo, Chrismo, Vivo, and Luxo. After a fast start, production of near beer fell, meaning less plant space was used throughout the thirteen dry years.

Many brewers manufactured ice cream, malt syrup, malt extract, and soda pop. Schlitz built the "E-Line" plant to make candies and chocolates. Anheuser-Busch shipped "high grade packing house products"—in other words, meat and yeast products.[123] Valentine Blatz distilled industrial alcohol. The David Stevenson Brewing Co. of Chicago produced spaghetti and macaroni.[124] Hamm's of St. Paul diversified with near beer, industrial alcohol, malt syrups, confectioner's supplies, soft drinks, sardines, cigars, and corncob pipes.[125] Griesedick Western of Belleville, Illinois, made ice and root beer.

In St. Louis, Griesedieck Beverage produced soft drinks and near beer, and processed hams and bacon. Coors of Golden, Colorado, became the world's third largest producer of malted milk. Olympia of Tumwater, Washington, made fruit-based soft drinks, Loju and Appleju, briefly. The Tumwater brewery was used for paper production, and then for cabinet making by a metalcraft company. Pabst produced near beer and malt syrups, but uniquely added cheese and metal pipe fitting manufacture.[126] One New York brewery went into the fur storage business.[127]

Cold Spring Brewing Company of Cold Spring, Minnesota, marketed mineral water during Prohibition, as it had when founded in 1874.[128] August Schell Brewing Company of New Ulm, Minnesota, made near beer that apparently had a poor taste when mixed with alcohol. It also sold root beer and cola, and wholesaled candy.[129] Incidentally, in the late 1970s, Schell used waste heat from its boilers to grow commercial quantities of tomatoes hydroponically.[130] This was a unique and unexpected sideline business for a brewery.

Evidence of specialized architectural modifications for these Prohibition-era businesses might appear in the archaeological record. For many breweries, the great change was that brewing equipment was removed. A spokesman for the company dismantling the Muehlebach brewery in Kansas City in 1930 said, "Unless the Mexicans want some of the machinery for [their breweries], we will put the metal and parts that have any present day use back into commercial activity, while the rest will go to the melting pot to be made into something else."[131]

Saloons

Many breweries, especially small ones, had saloons attached or nearby to retail their beer. In many instances, although these are poorly documented in the historical record, it was actually a case of a saloon doing small-scale brewing for its customers. Frequently, this was more like sharing homebrew than conducting a brewing business.

A noteworthy bartender's guide from 1862, as the lager revolution was in its adolescence, had instructions for saloons to properly serve the new style of beer:

> The refreshing qualities and flavor of Lager beer depends very largely on the manner of keeping and handling. Casks or kegs containing it should be kept at a temperature of about 40 degrees. Lager is always in its best condition when it comes from the brewer's ice-house. When carted through the streets on a hot summer's day, the temperature is quickly increased, and it must be stored in a refrigerator for three or four days in order to reduce it to a proper temperature before using.
>
> When the consumption of beer is sufficiently rapid, it is best drawn directly from the keg, the first glass drawn being rejected. The tap must be thoroughly cleansed before using; and as soon as the beer ceases to run freely, a vent is placed in the bung.
>
> When, however, the keg has to stand in use for some time before it becomes empty, a considerable amount of gas will escape every time the vent is opened, and the beer will become "flat, stale and unprofitable" at least for the consumer and retailer. To obviate this, and to keep the beer tolerably fresh to the end, the vent is not used, but is inserted in the vent-hole, leading to a receiver or cylinder containing air, compressed either by water-power or hand force pump. This exerts a continual pressure on the surface of the beer, and prevents the gas from rising. Too great an amount of air-pressure should be avoided, because the beer will be driven too forcibly through the tap, and fill the glass with more froth and less beer than a thirsty drinker would care to pay for.[132]

The unprecedented optimism, growth, and investment in the brewing business during the 1880s led to excess capacity in the 1890s. Brewers dropped their price per barrel of beer or offered special deals to particular customers. The depression of the early 1890s caused beer consumption to plummet over a million barrels between 1893 and 1894.[133] Beer taxes were soon raised to finance the Spanish-American War. If this was not enough, temperance forces had laws passed in many places, requiring very expensive licenses for saloons. Their plan was to limit the number of liquor outlets, and this would magically limit the alcohol consumption of society. Brewers found that many of their retail outlets could no longer afford their licenses. Brewers who were still financially sound advanced money for the licenses or purchased saloons, which they leased to operators. Competition was fierce and profit margins minuscule. These saloons sold the owner's beer exclusively and advertised it exclusively. This made some saloons virtual appendages of their brewery-owner. Excavation of a saloon or a brewery of the period must keep this connection in mind.

Remodeled Breweries and Entirely New Uses of the Structures

Preserving buildings after their main function and original reason for existence has ended is a challenge. Old breweries have become museums, office centers, and apartment complexes, among other things. Barry Cunliffe has mentioned that "historical archaeology often interfaces with the heritage industry and there is thus a need to consider conservation, reconstruction, management and presentation all of which raise quite difficult technical and ethical issues—issues of authenticity, the need for unbiased story lines and the all-too-frequent conflict between preservation and redevelopment . . ."[134]

The Schade Brewery on Trent Avenue in Spokane, Washington, makes an excellent example of reuse of a brewery. Architectural students at the University of Idaho about 80 miles away studied it extensively, and have written three architectural theses on their plans for restoration.[135] The brewery, which began operation in April, 1903, to brew 40,000 barrels per year, was a landmark in Spokane. Owner Paul Gustave Bernhardt Schade died in 1921, and after Prohibition, the brewery reopened under the name of Golden Age Breweries, Inc.[136] In 1957, the brewery was sold to the Bohemian Club Brewery and operated as their second plant. After Atlantic Brewing of Chicago bought the whole Bohemian Club Company, the plant was sold to Inland Metals Company, a scrap metal firm, in 1959. In 1984, the owner was not interested in doing any work on the building.

The students' evaluations of the building were very instructive to anyone doing archaeology or any type of analysis of a brewery of that era, size, and construction type. Original architectural drawings were located and used. The building offered a number of advantages and challenges. The brew house tower had large circular holes in the floor that were left when the tanks were removed. Pipes had all been removed, and in fact, "no mechanical systems" remained in the building.[137] Bearing walls were 3 feet thick in the basement and 2 feet thick on the third floor. Many windows had been filled in with brick. At the back of the building were keg storage rooms added in the 1940s. The boiler house was destroyed in 1968. The pasteurization room had a 40-foot arched window. One problem was the upper story, where wooden columns supported the roof trusses. Most of the building was concrete posts and integral beam slab. The elevator was operational but unsafe.

One architect proposed a restaurant/bar and a small hotel in the building. Another thought shops and offices were suitable. By December, 1992, a Schade Brewery Antique Mall was in place at the location, so some restoration did take place. Eventually, office space for many companies was made available. The building remains a Spokane landmark.

As of 2007, Portneuf Brewery in Pocatello, Idaho, was located in part of the old brewery once used by the East Idaho or

Figure 2.15. The modern Portneuf Brewery of Pocatello, Idaho, is now in the building of the old East Idaho Brewing Company, which closed in the 1950s. (Photo by author.)

Figure 2.16. New equipment being installed in Nampa, Idaho, after Prohibition. Owner Jake Lockman and the famous brewer Joseph Pickett are in the photo. (Photo from author's collection.)

Aero Club Brewery and then left unoccupied for decades. This is a case of a brewery replacing a brewery (fig. 2.15).

In 1986, Mervin Eisel planned to open a Minnesota brewing museum. He bought a brew kettle, capping machine, malt mill, bottle labeler, and machine for putting iron bands on wooden barrels from the defunct Schroeder Brewery in Perham, Minnesota. At about that same time, the old Figer's Brewery in Duluth had been transformed into the home of a shopping complex, restaurants, and a hotel.[138]

The Weiser, Idaho, brewery became a feed store, although it was briefly a bordello after its brewing days. The Weisgerber Brewery, Lewiston, Idaho, was used as a grocery warehouse before it burned one icy night in the 1930s. The Uniontown, Washington, brewery, which closed before Prohibition, became a bakery after 2000.

The Crescent Brewery in Nampa, Idaho, was demolished in the 1960s as part of a beautification program for the city (fig. 2.16). The magnificent Germanic castle that once brewed beer was replaced with a metal scrap yard. The impressive Lockman home beside it became offices as of 2006, which retain historic integrity.

The showpiece of the old Pearl Brewery, San Antonio, Texas, was the elliptical brick stable built in 1884.[139] When horses became passé, it was used for storage, then as an auditorium, and in 1971, as a restaurant. On permanent display in the courtyard is a locomotive engine that hauled materials into the brewery from 1947 to the 1970s.

Author Jerry Apps provided an extensive listing of the second lives of Wisconsin breweries as of the date of his book in 1992. A few of these follow. Ahnapee Brewing in Algoma became the Von Stiehl Winery.[140] Antigo Brewery in Antigo was used for a Midland Co-op store. The Citizens Brewery of Antigo became part of the Kraft plant. The Muench Brewer, Appleton, housed a mini shopping center. Joseph Hussa Brewery, Bangor, was converted to a feed mill. The Effinger Brewery in Baraboo took on new life as the home of the Circus World Museum. Beaver Dam Brewery in Beaver Dam became a mushroom factory. Henry Macke Brewery in British Hollow was a private residence. Cecil Brewery, Cecil, also became a feed mill. Cedarburg Brewery in Cedarburg housed an artist's collective. Oderbolz in Black River Falls was used for storage.

John Kurth and Company Brewery in Columbus housed a tavern. Denmark Brewery, Denmark became the print shop for a newspaper. Fox Lake Brewery, Fox Lake, had a bait shop on the ground floor and Dawes Laboratories upstairs. Breunig Brewery, Jefferson, had a tavern, electric motor repair shop, barber shop, and on the top floor was a residence. William Rahr and Sons Brewery, Manitowoc, was part of the Anheuser-Busch malt plant. Old Port Brewing Company, Port Washington, housed the American Legion Hall and bar.

In a large-scale project, the Valentine Blatz brewery in Milwaukee became an $8.8 million housing cooperative in 1983. The "hulking brick" was converted into 146 one- and two-story dwellings. It was hoped to be a model for other such projects around the country, and to bring people back into the city core.[141] The potential in sturdy old buildings is obviously diverse and great.

Malting and Hop Processing Facilities in Locations Separate from Breweries

Malting has been discussed as part of the brewing process, as have the technological changes that took place through the years. Another aspect to the malting business remains. Many breweries did not make their own malt but bought it from malting companies. The U.S. Census even lumped together the professions of brewer and maltster in recording occupation statistics during the nineteenth century. Malting companies also sold malt to distilleries that used it to make the mash that they distilled, and may have had a few other customers, but mostly sold to breweries.

Even in the twenty-first century, brewers with large malting facilities may buy special malts, such as those very darkly roasted and used in small amounts, from a malting company. Other brewers at this time have malting facilities far from any brewery they own. These are an integral part of the production of beer. For instance, Anheuser-Busch has operated a malting facility near Idaho Falls, Idaho, since 1991.[142] It ships about 8 million bushels of malt per year, working 24 hours a day, 365 days per year. From there, the malt goes to its breweries throughout the United States.[143] The company has no brewery in Idaho or any adjacent state.

Adjacent to this malting plant is the plant of the Groupo Modello Brewery of Mexico (of which Anheuser-Busch owns a significant portion). Idaho barley is malted there, shipped to Mexico, brewed into beer south of the border, and then much of the beer is imported into the United States.

Many nineteenth century malting stories from the eastern United States are known. In 1861, the first malting facility in Louisville, the Kentucky Malting Company, went into operation. By 1881, the malt houses of the city had $89,000 invested and twenty-six employees working 14-hour days to produce 680,00 bushels worth $510,000. Some rye and some corn malt was produced for distilleries.[144] The interconnected nature of the malting companies was obvious. For instance, the Kentucky Malting Company merged with the Stein and Doern malt house and became the Louisville Malt House with E. W. Herman as president and J. H. Pank as secretary/treasurer. Herman and Pank had a second plant in Louisville, but also owned the Chicago-based malting company called J. H. Pank and Company. Herman was also the president of a Knoxville, Tennessee brewery.[145]

Ferdinand F. Lutz had another Louisville malting company, but he eventually gave up the company and became an agent for the William Rahr and Sons Malting Company of Minneapolis and Manitowoc. By 1891, Kentucky was growing very little barley. Guetig and Sele, in their book on Louisville breweries, believed that brewers began to buy malt from these companies and use the space freed up at their breweries to house new refrigeration equipment.[146]

In the 1870s the largest grain market in the world was in Chicago. Chicago had numerous breweries and nearby Peoria was a center for distilleries, so malt was in high demand.[147] In addition, the many interconnected railroad lines made transportation easy and gave the maltsters an edge over eastern competition.

Chicago had twenty private malt houses by the mid-1880s. In 1882, Louis C. Huck, son of pioneer brewer John Huck, patented a system for air-drying and ventilating malt houses, and he also added a Saladin malting system to his malt house. In 1879, the malting trade in Chicago was valued at $1.5 million.[148] Also in 1879, Anheuser-Busch, despite having large malting facilities of its own, had to buy malt from two St. Louis malting firms.[149] The malt house Anheuser-Busch later built at the brewery still stands and has been converted to office space.

A book on the history of breweries in Buffalo, New York, found difficulty in separating malting and brewing industries because of the close connection.[150] As of 1996, three examples of nonaffiliated malting houses still stood in Buffalo.

In 1952—the 106th year of the company's existence—the Rahr Malting Company published a book with many photos and a small amount of text explaining its malting business.[151] The production chart began with a field of barley, then the grain went to a country elevator and on to the grain exchange. From there, it went to Rahr, either on a train car which could be dumped in one motion, or on a ship that was unloaded by the company's marine tower which was situated in Lake Michigan and enabled ships to come close enough for unloading without docking. Rahr then weighed the barley and put it in its storage elevators until needed. It was then cleaned, graded, and stored again.

The actual malting began with steeping. Next came germination, kilning, aging, blending, malt cleaning, loading, and then the malt was out the door to the buyer. A white-coated laboratory scientist appeared in the middle of the actual malting process, checking on every detail—at least on the flow chart in the book.

At the time of the book, Minneapolis was the world's largest barley exchange, and Rahr defiantly extolled its own modern, automated, efficient system of production. Rahr was particularly proud of its marine tower in Manitowoc with its giant bucket leg that could unload 18,000 bushels of barley per hour. From there, a belt conveyor carried the barley three city blocks to the "Head House." Barley from Canada or any location around the Great Lakes could reach the tower on the long cargo ships that plied the waters of the lakes. Barley that arrived by rail could be emptied at the rate of one car every six minutes by a giant device that lifted one car at a time. The barley was weighed on carload-size scales.

On the monitor floor the barley underwent light and heavy screenings, and then air stream and agitating screens to remove broken grains and stray seeds. Three floors of ring graders separated the grains into berries of similar size. Smaller berries were removed and sold for feed at a heavy discount. Each distinct nativity and variety of barley was binned separately according to berry size. Enormous storage and bin capacity made possible the "controlled and balanced blending which is the keystone of modern malting operations."[152]

The steeping required for the barley varied according to the variety, nativity, and berry size of the barley. Rahr solved this problem by having an abundance of steep tanks so that each type of barley could be soaked properly. After soaking, the barley went via overhead conveyors to the germination section. Each compartment there held 5,000 bushels. Giant helixes, like mechanical corkscrews, turned the growing barley and prevented matting. In older, smaller operations, wooden shovels and rakes did this job.

The barley was kept in germination for six days in precise temperature, humidity, and motion. First, tiny rootlets sprouted, then the acrospire started to grow through the kernel. This released the enzymes and modified the starch body. The perforated floors allowed uniform airflow, and the tilting tray allowed them to be emptied mechanically into the hopper below. The compartment was then cleaned with high-pressure water streams and was ready for the next load.

The grain then went to the kiln, where the application of heat stopped the growth of the green malt. Rahr had conventionally designed double-deck kilns and unique single deck kilns with airflow controls. The heat in 1952 was provided by steam, natural gas, or oil, or in any combination. Kiln fans that moved air at a rate of 160,000 cubic feet per minute were operated by variable electronic speed controls. A battery of instruments measured temperature and humidity at strategic points.

From the kiln, the malt was cooled, and then went to storage on a conveyor belt to be aged. Before malt reached a waiting train car, it was cleaned one more time, and "probed samples are taken according to prescribed Federal sampling procedures, hermetically sealed and filed."[153] Rahr claimed that sixty years before (1892), it had established the first scientific laboratory in an American malt house.

Structures Related to Hop Growing and Processing

Hops must be dried as soon after picking as possible. Like the other aspects of brewing, the hops industry has evolved greatly since the early brewing days in the United States.

On moderately sized hop farms in New England and New York, early simple kilns were made of stone and banked into the side of a hill. After the middle of the nineteenth century, frame kilns were more commonplace.[154] Early versions of these kilns had open fires fed with charcoal, with the hops on slats above them. Cast-iron stoves eventually replaced open fires. The next evolution came in the form of furnaces that forced hot air through the hops.[155] In time, oil-fired or gas-fired equipment replaced these. Hops were typically spread up to two feet deep for drying. Sulphur was sometimes burned in the kiln under the hops to bleach it uniformly and kill bacteria and molds.[156]

One kiln design recommended in New England had stone walls in a funnel shape, measuring 12 feet square at the top, 2 feet square at the bottom, and 8 feet deep. On top of the walls were joists, and on them lay a drying floor of lath covered with linen that was nailed down.[157] Over this, in later years, hop growers built a simple four-sided wooden structure. By making this house large, the hop grower gave himself room to store the dried hops, and by keeping the storage room from allowing steam to enter, the hops were kept in good condition.

The term "hop house" began to refer to both the kiln and cooling room, except later in Washington Territory where these were known as hop driers or dry houses. Stone foundations, a slatted drying floor, and charcoal fire were the standard ingredients. As Tomlan points out, directions for drying hops given to farmers did not change from 1801 to 1834, indicating little advancement in design concepts.[158]

A hop house from Maine built about 1880 is extant and two photographs appear in Tomlan's book on hops.[159] It was built on a hill and appears to be a single story from the front, where a large door is capable of admitting a wagon. The opposite end appears to be two stories with a porch over

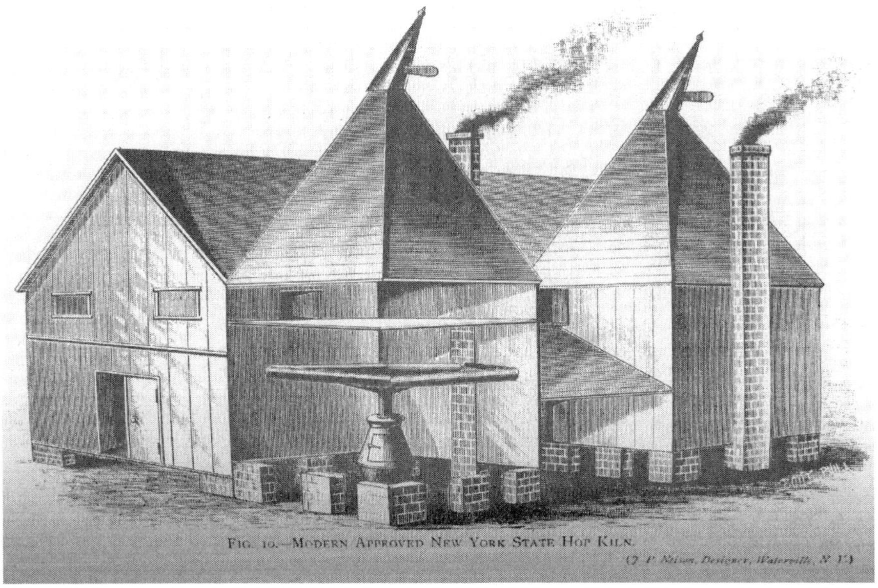

Figure 2.17. A New York hop kiln. (Meeker, 1883, 94–95.)

the lower level. The building was sided with shakes. From the exterior, it looks like a farm shed of no special purpose.

A good description exists of a hop house from Otsego County, New York: "An iron stove in the drying house heated hops on the upper, slatted floor. Airtight walls, a roof vent, and the wind-directed cowl above created the necessary updraft. Dried hops cooled in the top floor of the shed, . . . and were dropped to the press room for baling. The structure was painted 'barn red' with iron oxide and oil."[160] Another description of a typical hop house read: "A kiln floor 16 feet square, and laid four feet deep, will have about nine hundred bushels or 4,500 pounds of hops, which contain 3,400 pounds of water to be evaporated in twelve hours"[161] (fig. 2.17).

Ezra Leland described the hop kilns in his area of Morrisville in Madison County, New York, in 1845.[162] They had developed since the concepts published in 1834. They were about 30 feet long, 18 feet wide, and the first story of stone or brick was 9 feet high with 18-inch holes to let air in for the eight charcoal fires. The kiln had thinner walls than those of a decade earlier, and no funnel-shaped interior.

The next development in the eastern United States was the round stone oast house based on the British model. It was near the storage barn and would have a hallway to connect them. In 1832, balloon construction, using framing with two-by-fours, was introduced, and by 1860 was applied to hop kiln construction in Otsego County, New York.[163]

As hop growing moved to the Midwest, hop house construction developed there, starting with the concepts of the 1860s. H. H. Potter of Sauk County, Wisconsin, described the hop house he erected in 1865: "It was 40 by 20 with studding sixteen feet high under a single roof at quarter pitch which cost $400 to construct."[164] The hop kiln built by the Canfield family of Sauk County, Wisconsin, was an eight-sided log structure with a shake roof and a large ventilator on top. Others in the area were stone-and-rubble wall oast style. Tomlan concluded that "the most important contributions of Wisconsin's comparatively brief hop-growing effort lie in the introduction of the balloon frame and in the preoccupation with mechanical drying. A little over a decade later both of these improvements were adopted by the large growers on the Pacific Coast."[165]

Rectangular hop kilns were the only ones to become popular in the Pacific Northwest. In the northwest there were very large triple kiln hop houses with ramps on the side permitting horse-drawn wagons to unload.[166] The Woolrey-Koehler hop kiln of the Puyallup Valley of Washington state was built in 1869 of charred cedar logs. Wood stoves heated the entire building, including the upper level drying room and the central baling room (fig. 2.18). In 1890, the building was expanded and a second, shorter chimney added.[167]

After cooling, the hops were baled in America, while in Europe they were put in bags before being cooled. Hop baling presses were "invented by Harris of Waterville, New York, about the middle of the [nineteenth] century, and improved by Beardsley"[168] (fig. 2.19).

In 1867, William M. Haynie of Sacramento described the hop houses from his area as four-room frame structures under one roof, sometimes built by tenant Chinese farmers.[169] Later, some California hop kilns had wire screening instead of cloth to hold the hops, and soon kilns with two floors inside improved efficiency (fig. 2.20).

Figure 2.18. Wood stove for a hop kiln. (Meeker, 1883, 52.)

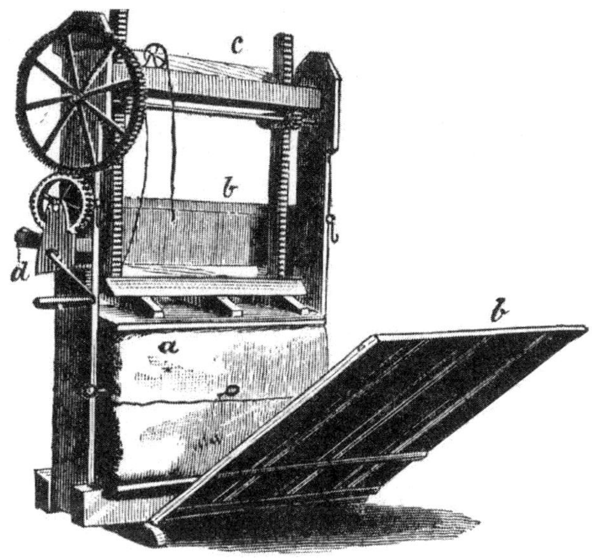

Figure 2.19. McCabe hop press. (Meeker, 1883, 36.)

Figure 2.20. Hop kilns of the Meeker Company in Puyallup, Washington. (Meeker, 1883, 26, 27.)

Figure 2.21. Coal stove for a hop kiln. (Meeker, 1883, 54.)

Washington hop growers seemed to vie to build the most ornate structures. J. R. Dickinson of Sumner, Washington, had a hop house measuring 96 by 44 feet and accented with an octagonal cupola on each ventilator, capped with a decorative spire.[170]

By the mid-1880s, steam-powered fans were attached to the hop house to force the hot air through the hops (fig. 2.21). In the 1930s, as the hop growers tried to shake off the effects of Prohibition, large electric fans became common. After World War II, corrugated steel became common for exterior wall sheathing.

As Kathleen Cooler wrote: "Memories [of old hop growing days] linger, however, and the aging, old-fashioned kilns that may still be seen near the traditional centers of hop production are nostalgic reminders of the days when the fields were bustling with life and cries of 'Wire down!' and 'Box full!' filled the air."[171] The history of hop agricultural structures certainly establishes their complexity and cultural development over time.

NOTES

1. Appel, "The Midwestern Brewery Before Prohibition: Development of an American Industrial Building Type," 57.
2. In October, 1873, George Gooding fell 12 feet from the landing at the top of the stairs at his brewery in Centerville, Idaho, and died two days later. *The Walla Walla Statesman* (Walla Walla, Washington), October 18, 1873, p. 2, c. 5. Three men fell through the floor of the malt house of the Koeninger Brewery in Ketchum, Idaho, in July 1884. *Ketchum Keystone* (Ketchum, Idaho), July 16, 1884.
3. Bob Gaynor, "The Breweries of Burlington, New Jersey," *American Breweriana Journal* 137 (September-October 2005), 33.
4. Appel, "The Midwestern Brewery Before Prohibition: Development of an American Industrial Building Type," 58.
5. Appel, "The Midwestern Brewery Before Prohibition: Development of an American Industrial Building Type," 58.
6. Appel, "The Midwestern Brewery Before Prohibition: Development of an American Industrial Building Type," 58.
7. Appel, "The Midwestern Brewery Before Prohibition: Development of an American Industrial Building Type," 57.
8. *One Hundred Years of Brewing*, 1903, 195.
9. Appel, "The Midwestern Brewery Before Prohibition: Development of an American Industrial Building Type," 65.
10. Appel, "The Midwestern Brewery Before Prohibition: Development of an American Industrial Building Type," 74.
11. *One Hundred Years of Brewing*, 1903, 135.
12. Appel, "The Midwestern Brewery Before Prohibition: Development of an American Industrial Building Type," 76–82.
13. Appel, "The Midwestern Brewery Before Prohibition: Development of an American Industrial Building Type," 77.
14. *One Hundred Years of Brewing*, 1903, 136.
15. Appel, "The Midwestern Brewery Before Prohibition: Development of an American Industrial Building Type," 57.
16. *One Hundred Years of Brewing*, 1903, 135.
17. Appel, "The Midwestern Brewery Before Prohibition: Development of an American Industrial Building Type," 229.
18. Apps, *Breweries of Wisconsin*, 177.
19. Appel, "The Midwestern Brewery Before Prohibition: Development of an American Industrial Building Type," 118.
20. Appel, "The Midwestern Brewery Before Prohibition: Development of an American Industrial Building Type," 120.
21. Appel, "The Midwestern Brewery Before Prohibition: Development of an American Industrial Building Type," 121.
22. Appel, "The Midwestern Brewery Before Prohibition: Development of an American Industrial Building Type," 135.
23. Appel, "The Midwestern Brewery Before Prohibition: Development of an American Industrial Building Type," 162–163.

24. *One Hundred Years of Brewing*, 1903, 137. The back files of *The Western Brewer* covered and documented this whole era of brewery construction according to this source.
25. Scammel and Colyer, *Breweries and Maltings: Their Arrangement, Construction, Machinery, and Plant*, 1880, 5.
26. Appel, "The Midwestern Brewery Before Prohibition: Development of an American Industrial Building Type," 193.
27. Appel, "The Midwestern Brewery Before Prohibition: Development of an American Industrial Building Type," 184.
28. Appel, "The Midwestern Brewery Before Prohibition: Development of an American Industrial Building Type," 184.
29. Appel, "The Midwestern Brewery Before Prohibition: Development of an American Industrial Building Type," 185.
30. Skilnik, *The History of Beer and Brewing in Chicago, 1833–1978*, 33.
31. Holian, *Over the Barrel: The Brewing History and Beer Culture of Cincinnati, 1800 to the Present. Volume One (1800-1919)*, 124.
32. *One Hundred Years of Brewing*, 1903, 145.
33. Appel, "The Midwestern Brewery Before Prohibition: Development of an American Industrial Building Type," 274–275.
34. Appel, "The Midwestern Brewery Before Prohibition: Development of an American Industrial Building Type," 274–275.
35. Appel, "The Midwestern Brewery Before Prohibition: Development of an American Industrial Building Type," 278.
36. Appel, "The Midwestern Brewery Before Prohibition: Development of an American Industrial Building Type," 278.
37. Appel, "The Midwestern Brewery Before Prohibition: Development of an American Industrial Building Type," 307.
38. Franklin, Hayes, and Fleigner of Pocatello, Idaho, had their brewery designed by Louis Lehle of Chicago. This shows how far his influence spread. *Pocatello Tribune*, (Pocatello, Idaho), January 22, 1902.
39. Susan K. Appel, "Chicago and the Rise of Brewery Architecture," *Chicago History* (Spring 1995), 4–19. Wolf and Lehle were associated for about ten years.
40. Appel, "New Finds in Brewery Architecture," 7.
41. Skilnik, *The History of Beer and Brewing in Chicago, 1833–1978*, 31.
42. Skilnik, *The History of Beer and Brewing in Chicago, 1833–1978*, 32.
43. Kious and Roussin, "The Breweries of Collinsville, Il.," 22.
44. Kious and Roussin, "The Breweries of Collinsville, Il.," 23.
45. Kious and Roussin, "The Breweries of Collinsville, Il.," 23.
46. Appel, "The Midwestern Brewery Before Prohibition: Development of an American Industrial Building Type," 201.
47. Appel, "The Midwestern Brewery Before Prohibition: Development of an American Industrial Building Type," 202.
48. Scammel and Colyer, *Breweries and Maltings: Their Arrangement, Construction, Machinery, and Plant*, 88.
49. Vance, *Beer in the Beehive: A History of Brewing in Utah*, 79.
50. Ronnenberg, "Idaho on the Rocks: The Ice Business in the Gem State," 5.
51. Ronnenberg, "Idaho on the Rocks: The Ice Business in the Gem State," 5.
52. *Idaho Daily Statesman* (Boise, Idaho), September 2, 1891, p. 8, c. 1.
53. Guetig and Selle, *Louisville Breweries: A History of the Brewing Industry in Louisville, Kentucky, New Albany and Jeffersonville, Indiana*, 136–137.
54. Millard, *History of the Carson Brewing Company*, 7.
55. *One Hundred Years of Brewing*, 1903, 445.
56. "The Story of Kessler Brewery, Established 1865 and How Beer is Brewed" (Helena, Montana, n.d., c.1952), 4.
57. *Idaho Tri-Weekly Statesman* (Boise, Idaho), July 11, 1868, p. 3, c. 2.
58. *Wood River Times* (Weekly), (Hailey, Idaho), February 11, 1891, p. 1, c. 4.
59. Challis Brewery Historic District, United States Department of the Interior, National Register of Historic Places Inventory, Nomination Form.
60. *Idaho County Free Press* (Grangeville, Idaho), July 28, 1893, p. 1, c. 3.
61. *Idaho County Free Press* (Grangeville, Idaho), January 6, 1893, p. 1, c. 6.
62. *Idaho County Free Press* (Grangeville, Idaho), July 7, 1893, p. 1, c. 5.
63. *Idaho County Free Press* (Grangeville, Idaho), August 18, 1893, p. 1, c. 6.
64. *Idaho County Free Press* (Grangeville, Idaho), September 1, 1893, p. 1, c. 4.
65. *Idaho County Free Press* (Grangeville, Idaho), September 8, 1893, p. 1, c. 6.
66. *Idaho County Free Press* (Grangeville, Idaho), October 13, 1893, p. 1, c. 6, and p. 1, c. 5.
67. *Idaho County Free Press* (Grangeville, Idaho), December 8, 1893, p. 1, c. 4.
68. *American Brewer's Review* (n.p., n.p., 1909), 343.
69. *Idaho Register* (Idaho Falls), May 6, 1904.
70. Quoted in Dick, "A Geographical Analysis of the Development of the Brewing Industry of Minnesota," 227.
71. Appel, "The Midwestern Brewery Before Prohibition: Development of an American Industrial Building Type," 214.
72. Holian, *Over the Barrel: The Brewing History and Beer Culture of Cincinnati, 1800 to the Present. Volume One (1800-1919)*, 213.
73. Bill Yenn, *The American Brewery: From Colonial Evolution to Microbrew Revolution* (St. Paul, Minn.: MBI, 2003), p. 32.

74. Appel, "The Midwestern Brewery Before Prohibition: Development of an American Industrial Building Type," 88.
75. Miller, *Breweries of Cleveland*, 34–35.
76. Marlin R. Ingalls, "A Phase II Historic Architectural Evaluation of The Law Brewery Cave, Site 31-00008 Primary Roads Project STP-32-1 (1)—2C-32 a.k.a. PIN 95-31080-1 Dubuque County, Iowa." Office of Project Planning, Iowa Department of Transportation, Ames, Iowa 50010. Carl A. Merry and Stephen C. Lensink, Co-Principal Investigators, Project Completion Report Volume 22, Number 16, 1999, p. 11.
77. Salem, *Beer, Its History and Its Economic Value as a National Beverage*, 71.
78. Downard, William L. *The Cincinnati Brewing Industry, A Social and Economic History*, 31.
79. Kelly J. Dixon, *Boomtown Saloons: Archaeology and History in Virginia City* (Reno: University of Nevada Press, 2005), 26.
80. Sally Zanjani, *Glory Days in Goldfield, Nevada* (Reno: University of Nevada Press, 2002), 48–49.
81. Apps, *Breweries of Wisconsin*, 168.
82. Herman Ronnenberg, "Researching Your Hometown Brewery: The Fire Insurance Connection," *The Breweriana Collector* 69 (Spring, 1990): 21–24. The discussion of the fire maps comes from this source.
83. There are two important books about these maps; both can give the novice researcher the basic information to get started. R. Philip Hoehn, *Union List of Sanborn Fire Insurance Maps Held by Institutions in the United States and Canada* [2 vols.] (Santa Cruz, California: Western Association of Maps Libraries, 1976–1977) and Library of Congress, Geography and Map Division, Reference and Bibliography Section, *Fire Insurance Maps in the Library of Congress: Plans of North American Cities and Towns Produced by the Sanborn Map Company: A Checklist* (Washington, D.C.: Library of Congress, 1981). These sources have sections for each state, and within these, each city is listed along with the dates for all the editions of its maps. Hoehn's book also lists library holdings; the Library of Congress book lists only those held in that institution.
84. See the sidebar. I have compiled a list of Idaho brewery owners who worked with local fire departments. It was naturally in their self-interest to have fire protection, but their interests seems to have gone beyond this; they enjoyed being part of the institutions that protected their cities.
85. Dick, "A Geographical Analysis of the Development of the Brewing Industry of Minnesota," 93.
86. "The Story of Kessler Brewery, Established 1865 and How Beer is Brewed" (Helena, Montana, n.d., c.1952,) 10.
87. *The Western Brewer* 34:9 (September 1909), 472.
88. *Idaho Daily Statesman* (Boise, Idaho), August 7, 1909, p. 2, c. 2.
89. *Idaho Daily Statesman* (Boise, Idaho), August 7, 1909, p. 2, c. 2.
90. Hull-Walski and Walski, "Brewing and Bottling in Harpers Ferry, West Virginia," 17.52.
91. Appel, "The Midwestern Brewery Before Prohibition: Development of an American Industrial Building Type," 224.
92. Appel, "The Midwestern Brewery Before Prohibition: Development of an American Industrial Building Type," 227.
93. "The Year in Review," *The Boomerang* (Palouse, Washington), Winter, 1995: 10. See Herman Ronnenberg, "The Brewery of Palouse, Washington," *Bunchgrass Historian* 32, no. 1, (2006): 4–11. Herman Ronnenberg, "The Palouse Washington Brewery," *The Breweriana Collector* (Spring 2003): 21–22.
94. Appel, "The Midwestern Brewery Before Prohibition: Development of an American Industrial Building Type," 227.
95. *Idaho County Free Press* (Grangeville, Idaho), Dec. 4, 1880.
96. Judy D. Tordof, "Current Research: Pacific West: Nevada," *Society for Historical Archaeology Newsletter* 31, no. 3 (Fall 1998): 25.
97. *The Nez Perce News* (Lewiston, Idaho), April 20, 1882.
98. *Spokesman-Review* (Spokane, Washington), Oct. 11, 1905.
99. *Boise News* (Bannock City [Idaho City]), Dec 19, 1863.
100. *Idaho County Free Press* (Grangeville, Idaho), July 15, 1892.
101. Slade, "History of the Billings Brewing Company," 12.
102. Ingalls, "A Phase II Historic Architectural Evaluation of The Law Brewery Cave, Site 31-00008 Primary Roads Project STP-32-1 (1)—2C-32 a.k.a. PIN 95-31080-1 Dubuque County, Iowa," 8–10.
103. Ingalls, "A Phase II Historic Architectural Evaluation of The Law Brewery Cave, Site 31-00008 Primary Roads Project STP-32-1 (1)—2C-32 a.k.a. PIN 95-31080-1 Dubuque County, Iowa," 29.
104. Dick, "A Geographical Analysis of the Development of the Brewing Industry of Minnesota," 218.
105. Apps, *Breweries of Wisconsin*, 113.
106. Hubert Rother and Charlotte Rother, *Lost Caves of St. Louis* (St. Louis: Virginia Publishing Co., 1996), 25. The discussion of the St. Louis caves comes from this source unless otherwise noted.
107. Appel, "The Midwestern Brewery Before Prohibition: Development of an American Industrial Building Type," 82.
108. Appel, "The Midwestern Brewery Before Prohibition: Development of an American Industrial Building Type," 83.
109. Skilnik, *The History of Beer and Brewing in Chicago, 1833–1978*, 5.
110. Moira F. Harris, "From Limestone to Rubble at a Brewery Full of Memories," *American Breweriana Journal* 118:15.
111. John T. Flanagan, *Theodore Hamm in Minnesota: His Family and Brewery* (Minneapolis: Pogo Press, 1989), 14.
112. *Idaho Tri-Weekly Statesman* (Boise, Idaho), June 16, 1877, p. 3, c. 2.

113. Maxwell and Sullivan, *Hometown Beer: A History of Kansas City's Breweries*, 75.
114. Plavchan, "A History of Anheuser-Busch, 1852–1933," 81.
115. Skilnik, *The History of Beer and Brewing in Chicago, 1833–1978*, 9.
116. Maxwell and Sullivan, *Hometown Beer: A History of Kansas City's Breweries*, 75.
117. Maxwell and Sullivan, *Hometown Beer: A History of Kansas City's Breweries*, 93.
118. Elliot West, *The Saloon on the Rocky Mountain Mining Frontier* (Lincoln: University of Nebraska Press, 1979), 108–9.
119. Ronnenberg, *Beer and Brewing in the Inland Northwest*, 92.
120. Ronnenberg, *Beer and Brewing in the Inland Northwest*, 92.
121. Ronnenberg, *Beer and Brewing in the Inland Northwest*, 92–3.
122. Ronnenberg, "Idaho on the Rocks: The Ice Business in the Gem State," 5.
123. Plavchan, "A History of Anheuser-Busch, 1852–1933," 182.
124. Dick, "A Geographical Analysis of the Development of the Brewing Industry of Minnesota," 60.
125. Herman Ronnenberg, *The Politics of Assimilation* (New York: Carlton Press, 1975), 83.
126. Ronnenberg, *The Politics of Assimilation*, 83.
127. Stack, "Liquid Bread: An Examination of the American Brewing Industry, 1865–1940," 102.
128. Dick, "A Geographical Analysis of the Development of the Brewing Industry of Minnesota," 125.
129. Dick, "A Geographical Analysis of the Development of the Brewing Industry of Minnesota," 125–6.
130. Dick, "A Geographical Analysis of the Development of the Brewing Industry of Minnesota," 172.
131. Maxwell and Sullivan, *Hometown Beer: A History of Kansas City's Breweries*, 44.
132. Thomas, *The Bar-Tenders Guide, or How To Mix All Kinds of Plain and Fancy Drinks*, 17.
133. Miller, *Breweries of Cleveland*, 101.
134. Barry Cunliffe, "Afterward: Historical Archaeology in the Wider Discipline," in Dan Hicks and Mary C. Beaudry, eds., *The Cambridge Companion to Historical Archaeology* (Cambridge: Cambridge University Press, 2006), 318.
135. Gregg B. Creighton, "Schade Brewery Restoration: Preservation of a Prototype" (bachelor's thesis in architecture, University of Idaho, 1984). Judy R. Guse-Noritake, "The Schade Brewery, A Restoration" (bachelor's thesis in architecture, University of Idaho, 1978). Kristine M. Slaybaugh, "Schade Brewery Development" (thesis, College of Art and Architecture, University of Idaho, 1987).
136. During the early days of the Depression, the brewery was an unofficial hotel for the hobos from the nearby railroads and was called the "Hotel de Gink."
137. Creighton, "Schade Brewery Restoration: Preservation of a Prototype," 31.
138. Diane Mundt, "Professor Planning Museum of Minnesota Brewing," *Minneapolis Star and Tribune* (Minneapolis, MN), August 6, 1986, p. 4B.
139. Tom Rejmaniak, "What Ever Happened to the Pearl Brewery Stables?" *American Breweriana Journal* 144 (November-December 2006): 15.
140. Apps, *Breweries of Wisconsin*, 174. All the Wisconsin examples that follow are from Apps's book.
141. Dan Day, "Brewery Becoming Co-op Apartments," *The Spokesman-Review* (Spokane, Washington), February 13, 1983, p. E2.
142. Ground was broken in 1989, and the first malt was produced in 1991. Communication from Carl Adam, 2007.
143. During December 2006 and January 2007, the plant shipped malt to Los Angeles and Fairfield, CA; Williamsburg, VA; Jacksonville, FL; Cartersville, GA; Baldwinsville, NY; Merrimack, NH; Columbus, OH; and Newark, NJ. Correspondence with Carl Adam, February 2007.
144. Guetig and Selle, *Louisville Breweries: A History of the Brewing Industry in Louisville, Kentucky, New Albany and Jeffersonville, Indiana*, 29.
145. Guetig and Selle, *Louisville Breweries: A History of the Brewing Industry in Louisville, Kentucky, New Albany and Jeffersonville, Indiana*, 29–30.
146. Guetig and Selle, *Louisville Breweries: A History of the Brewing Industry in Louisville, Kentucky, New Albany and Jeffersonville, Indiana*, 30–31.
147. Skilnik, *The History of Beer and Brewing in Chicago, 1833–1978*, 30.
148. Skilnik, *The History of Beer and Brewing in Chicago, 1833–1978*, 37.
149. Plavchan, "A History of Anheuser-Busch, 1852–1933," 60.
150. Stephen R. Powell, *Rushing the Growler: A History of Brewing in Buffalo* (Buffalo, NY: Apogee Design, 1996), 109.
151. Hans Hinrichs, ed., *Rahr* (Manitowoc, Wisconsin: Thomas Delehanty Inc., 1952).
152. Hinrichs, *Rahr*, 1952.
153. Hinrichs, *Rahr*, 1952.
154. Tomlan, *Tinged with Gold: Hop Culture in the United States*, 66.
155. Tomlan, *Tinged with Gold: Hop Culture in the United States*, 68. Includes reproductions of advertisements for "The Granger," a coal-fired small round stove from 1878, and a cast iron corrugated hop furnace from 1935.
156. Cooler, "Hop Agriculture in Oregon: The First Century," 19.
157. Tomlan, *Tinged with Gold: Hop Culture in the United States*, 169.
158. Tomlan, *Tinged with Gold: Hop Culture in the United States*, 169.

159. Tomlan, *Tinged with Gold: Hop Culture in the United States*, 67.

160. *Historical Atlas of the United States* (Centennial Edition), (Des Moines, Iowa: National Geographic, 1988), 131. The illustration here shows a rather small but distinctive structure.

161. *One Hundred Years of Brewing*, 1903, 57.

162. Tomlan, *Tinged with Gold: Hop Culture in the United States*, 170.

163. Tomlan, *Tinged with Gold: Hop Culture in the United States*, 178.

164. Tomlan, *Tinged with Gold: Hop Culture in the United States*, 188.

165. Tomlan, *Tinged with Gold: Hop Culture in the United States*, 195.

166. Priscilla Tiller, "Hops: The Curious Crop of Lewis County," *Landmarks, Magazine of Northwest History and Preservation* 1, no. 2 (Winter 1982): 22.

167. Washington State Office of Archaeology and Historic Preservation, *Built in Washington, 12,000 Years of Pacific Northwest Archaeological Sites and Historic Buildings* (Seattle: Washington State University Press, 1989), 51.

168. *One Hundred Years of Brewing*, 1903, 57.

169. Tomlan, *Tinged with Gold: Hop Culture in the United States*, 196.

170. Tomlan, *Tinged with Gold: Hop Culture in the United States*, 200.

171. Cooler, "Hop Agriculture in Oregon: The First Century," 80.

CHAPTER THREE

Makers' Marks, Patents, Labels, Production Dates, Beer Containers, and Other Evidence

The widespread nature of brewing and beer consumption left behind countless objects that are meaningful to the history of beer in the United States. An entire culture has since evolved that revolves around these fairly mundane items, and this community has accomplished important work with regard to the history of brewing.

Breweriana Collecting Clubs

In 1971 the Beer Can Collectors of America organization was founded, followed by the Eastern Coast Breweriana Association that year. The National Association of Breweriana Advertising formed in 1972, and the American Breweriana Association (ABA) came along in 1982. In addition to socializing, these clubs published research on beer cans, brewery collectibles, and the breweries that produced them. The sometimes meticulous work of hundreds of collectors has created a volume of information that should not be overlooked by archaeologists. Books and articles exist on the dating and production of dozens of items that could be found in the archaeological record. Many members have worked in an industry related to brewing, or come from families that once owned breweries. Questions addressed to the membership will almost certainly produce an informed answer and offers of assistance. In 2008, the ABA successfully opened the National Brewery Museum and Library in the old Potosi Brewery in southwestern Wisconsin which offers a wealth of information.

Breweriana collectors generally specialize in items that might include ash trays, barrels, books and magazines, bottles, brewery equipment, business cards, calendars, cans, cases, chairs, clocks, coasters, corkscrews, crates, crowns, foam scrapers, glasses, history, knives, labels, lamps, leaded windows, lithographs, match safes, matches, medals, menus, mini beers, mirrors, mugs and steins, neons, openers, paper items, patches, photos, pinbacks, pitchers, playing cards, post cards, reverse on glass, salt shakers, signs, statues, tap knobs, thermometers, tip trays, tokens, trays, and watch fobs.

The value of the work done by breweriana collectors can be illustrated by a pamphlet published by Bob Kay of Illinois in 1993.[1] Starting with the first Budweiser label in 1878, Kay documented and illustrated all the changes up to the date of publication. More than one hundred variations have occurred in the label, not counting such things as bottle type, alcohol content, and so on. He also lists start-up dates for each of the Anheuser Busch branches because labels from each place differ. A Budweiser bottle with the label intact could be dated to within a few years with this source as a guide. This concept would, of course, generally be more valuable for dating sites with beer bottles than breweries themselves, but beer bottles and cans of stray types do find their way into breweries.

Ed Love and J. Peterson Powell wrote two volumes which each illustrate, describe, and date 239 trademarks for beers, breweries, and related products that have been registered with the U. S. Patent & Trademark Office.[2] This could help to identify anything made by a brewer that had a trademark on it, and the introduction helps explain the pitfalls of researching trademarks.

Since the 1970s, numerous books have appeared that catalog the beer cans made in the United States. Since the turn of the twenty-first century, several have been published that are noteworthy in their comprehensive, encyclopedic approach. Kevin C. Lilek published a mammoth volume of early beer cans followed by a large supplement based on thirty years of work.[3] He explains that early cans had an insert on the label explaining and illustrating the use of an opener to punch a hole in the can for the consumer to gain access, and are thus known as "opening instruction flat tops." He has photographs of virtually all of these. Three pictures are provided for each can so all sides are shown.

Metal Kegs

"In 1933 breweries changed over to metal rather than wooden beer barrels. Metal barrels required no upkeep other than washing and pitching. Wooden barrels had to be coopered and sprinkled with water to keep them tight."[4] These early metal barrels were galvanized (zinc coated) iron. Schaefer Brewing of New York City was still using wood kegs in 1941, as shown in the company's book commemorating its 100th year of operation. Thus, introduction dates vary.

By 1949, Reynolds Aluminum advertised its kegs in *Brewers' Digest*.[5] In the ad, a worker holds two kegs with apparent ease due to the light weight and "knuckle-deep chimes for a surer grip, the full extended rings to roll perfectly true on ramps. Full, double drainage on bottom, too . . ." The company proclaimed that over a quarter of a million barrels had been produced in the past two years. The advertising also explained that the "Exclusive Benson Bung Plate (U.S. Pat. No. 2460410) is the secret of barrel's aluminum interior. Barrel not pierced and absence of welding clusters insures smooth flow of beer, easy draining of cleaning fluid. Forged, welded bung plate offers tremendous resistance to impact."[6]

These were the Hoff-Stevens style kegs which remained popular from the 1940s to the 1970s and remained in use far beyond. They had two holes, one on the side for filling and one on top for tapping. Once filled, the side hole had a wooden bung pounded into it. By the 1980s, most large brewers changed to the sleeker Sankey keg, which was easier to clean, fill, and transport—and had no bung.[7]

Many of the older style kegs ended up in metal recycling facilities. In 1994, the new Yards Brewing Co. found that these were in junkyards and available for five to ten dollars each. Many other microbreweries began to use them as well. Many old Hoff-Stevens kegs had the original owner's name permanently affixed, which should be of note to an archaeologist or historian trying to identify them (fig. 3.1). By 2002, 95 percent of the kegs at the Great American Beer Festival in Denver were Sankey type.

In 2002, Schaefer Kegs of North America offered stainless steel, polyurethane, and polypropylene kegs in sizes from 5 liters to a half-barrel (58.6 liters). Its "Keggy" was a complete dispensing system with CO_2 tank, regulators, and keg all built in.[8]

Figure 3.1. Metal keg from the 1930s, Wallace, Idaho. The keg was labeled Hartz Brewery which had a distributor but not a brewery there. (Photo by author.)

Beer Bottles

Bottle collectors believe that the ale bottles used by colonial era brewers were made of pottery and imported from England.[9] The oldest ones had a matte or unglazed surface, and are very similar to gin bottles of the era, but easily distinguished as beer bottles. Casper Wistar in New Jersey started the first successful American glasshouse in 1739.[10] Perhaps he made some bottles, but certainly this would be the earliest an American-made glass bottle could be dated. Black beer bottles made in three-piece molds and rarely embossed were imported before the mid-nineteenth century.

Investigations at the Canterbury Shaker Village in central New Hampshire in the late 1990s unearthed dozens of stoneware beer and ale bottles from the mid-nineteenth century.[11] This led the researchers to question the degree of separation and purity the Shakers were seeking to maintain. This is an example of using beer bottles in site interpretation.

Bottled porter was imported into Louisville, Kentucky, by 1811, possibly from Pennsylvania.[12] In 1813, only 143 barrels and two hundred gross bottles of porter were shipped through the Falls of the Ohio. This was a minuscule amount compared to the whiskey, cider, and even peach brandy in commerce.

During the first thirty years of brewing in Wisconsin, from 1840 to 1870, pottery bottles with the names of the brewery stamped (debossed) on them were used.[13] Several different lip styles were known on these bottles, but all must have used corks for closures. Apparently these were manufactured in the general area where they were used. Likewise, in Minnesota, early brewers used pottery bottles with their names debossed.[14] Wayne Kroll cataloged the initials found on the bottom or base of bottles discovered in Wisconsin.[15] For instance, "CLYDE" on a bottle identifies it as originating from the Clyde Glass Works in Clyde, N.Y., which operated from 1827 to 1913. A glass plant that manufactured beer bottles, among other unrelated items, was founded in Milwaukee in 1880 and fared well until Prohibition.[16]

By 1840, Louisville had a glass works making porter bottles, among other items, and by 1849, the Kentucky Glass

Works opened.[17] By 1844 the Metcalf Brewery there advertised bottled beer.

In 1876, the San Francisco and Pacific Glass works merged, and bottles made on the Pacific Coast after that time have makers' marks of SFPGW on the bottoms.[18] George Newman formed the new Pacific Coast Glass Works about 1900 and the bottles had PCGW on the bottoms.

Before 1886, Adolphus Busch bought beer bottles from four companies: the Mississippi Glass Co. and Lindell Glass Co., both of St. Louis, the DeSteiger Glass Co. of La Salle, Illinois, and the Pittsburgh City Glass Co. of Pennsylvania.[19] By 1886, inadequate supplies led Busch to purchase the Belleville Glass Co. just across the Mississippi in Belleville, Illinois; he also purchased an interest in the Streator Bottle and Glass Co. of Streator, Illinois.[20] After research on both sides of the Atlantic, Busch built his own bottle plant, the Adolphus Busch Glass Manufacturing Co. in 1892. He supplied bottles for himself and St. Louis rival Lemp's Western Brewery.[21]

A photo from the Christian Moerlein Brewery in Cincinnati, dated to the early 1890s, shows employees cleaning brown glass bottles and Old Jug Lager crockery bottles interchangeably on the bottle washing machine.[22] A few pottery bottles remained in use for decades. In this instance, Moerlein apparently used this type of bottle as an advertising connection with the brand name, which they introduced in June, 1891.

Classifying Beer Bottles

Collectors divide beer bottles into four categories: porter bottles from 1820 to 1920; ale from 1845 to 1850; early lager from 1847 to 1850; and late lager from 1850 to 1860.[23] In the late 1860s breweries began to emboss bottles with their names and advertising messages. The Swan Brewery was one of many that used embossed bottles.

The manufacture of bottles can be roughly dated by the length of the seam on the side of the bottle, resulting from the mold in which the bottle was made. Before 1860, seams extended just over the shoulder of the bottle; the neck had no seam. From 1860 to 1880, the seam was longer, going most of the way up the neck. Between 1880 and 1890, the seam went farther up the neck to the lip, but not over it. From 1900 to the present, the seam runs the entire length of the bottle.[24] Before 1860, glass workers formed bottles by using a blowpipe and hand tools. After 1860 the glass was blown into a mold, which made the bottles more uniform. The lip was applied at the end of the process. Around 1880 the closed mold became common; in this process the neck and lip were formed mechanically.

Bottles that appear very old yet lack mold seams were probably made with a process called "turned in the mold," which created distinctive circular marks on the bottle. No embossing is possible on these bottles.

The lip of the bottle also gives clues as to its date of manufacture. The Standard Blob was manufactured from 1840 to 1910. From 1895 to 1919 three types are known: Applied Crown, Tapered Blob, and Trimmed Blob. The Machine Blob is dated from 1910 to 1920, and from 1910 to the present, the Crown Top is recognized.[25] These dates, as given by Wayne Kroll, are very general. The lips on the post-1910 bottles have seams that go to the very top.

Prior to 1840, bottles have a "sheared lip." The process of using shears to cut the bottle free of the blowpipe created this effect. About 1840 a glass ring termed a "laid-on-ring" was put around the sheared lip. Between 1840 and 1880, at least twelve different types of lips were made using different forming tools. They were variously rounded, flanged, plain, double rounded, and so on.[26]

At least six different finishes or lip styles were found at the Klondike Brewery in Dawson, Canada.[27] Most bottles were Crown for crown cap closures. All other bottles recovered were designed for cork closures: beaded lager for cork closure; flattened side for cork closure—the only type with a one-part rather than two-part finish; champagne-style for cork closure; flattened side with V-shaped bead for cork closure; and castor oil with bead for cork closure.

Some bottles are described as black glass open pontil. The glassblower used a pontil, or iron rod, to hold the still molten bottle so a top could be attached. The pontil was then broken off, leaving a scar. These generally date up to 1860.[28] Taper tops (sometimes called taper blobs) were made in a snap case, which alleviated the need for a pontil. They are usually squat quart bottles with a ring below the tapered top. Taper tops were made from approximately 1860 to 1880.

Some bottles were manufactured even before the 1800s with threads inside the necks; these were specialty bottles and very difficult to manufacture. After 1902 threads could be put on the outside of the lip but these did not become common until 1924. An embossed Dumfries Ale bottle from England with inside threads in the neck is known. A trend to make whiskey bottles like this started about 1875. The bottle used a rubber stopper, but the design was a short-lived failure.[29]

Imperfections and blemishes in the glass also give a clue to the age of a bottle. Air bubbles were trapped in the glass until about 1920. Bottle sides were often of uneven thickness in older bottles, and marks from the mold might show on the outside of a bottle. The color of a bottle is also of note. Before 1840, both intentionally colored glass and intentionally clear glass were rare specialty items. Until 1880 or so, the color of the glass for mass-market items was considered unimportant. Iron slag used in glass up to 1860 produced a dark olive green or olive amber glass that has become known as "black glass."[30] In 1900, for instance, Anheuser-Busch sold Budweiser and Faust in clear bottles, Standard Anheuser and Pale Lager in amber bottles, and Export was in bottles of either color.[31]

For identifying bottles, nothing is better than embossing if it is present. Embossing, or raised letters manufactured right into the glass, was used for identifying and advertising the product and the brewery. Glass manufacturers had two ways of producing embossed bottles. They could manufacture an entirely new mold for the customer, or they could use an existing mold and change the plate to make the impression, a less expensive alternative. Plate molds, or "slug plates," left a round impression on the bottle with the lettering inside.

Bottles with the embossing on the center are older and more eagerly sought. When paper labels became common, this location was in the way. Bottles could also be embossed on the base, shoulder, or bottom. These alternatives did not interfere with a paper label being attached. As brewers had separate brands of beer to promote, these manufacturing alternatives took over.

Another way to put information onto a bottle was to paint or enamel a label. This began during the 1930s, became fairly popular in the 1940s, and was less used after that time.[32] As with many variations in beer bottles, a few contemporary examples are known; old ideas never die—they merely hibernate and reappear in special circumstances.

Bottles in the Archaeological Record

One problem with identifying archaeological sites and dates from beer bottles is that breweries would use the bottles of other brewers. For instance, in the assemblage of bottles found at the Klondike Brewery in Dawson, Yukon Territory, were twenty-six from the Adolphus Busch Glass Manufacturing Co. of Belleville, IL, and St. Louis, MO.[33] Identification of bottles by their base markings indicated that four were found in one pit and ninety-seven in another from the American Bottle Co. of Chicago or Toledo; two were from San Francisco; three arrived from Terre Haute, Indiana; and nineteen originated from Wooster, Ohio. Anheuser-Busch shipped 1,290,744 bottles of beer to the Fairbanks, Alaska, Northern Commercial Company between 1911 and 1917.[34] Probably, none were ever returned to St. Louis, but many were reused regionally.

Despite difficulties, brewers in Idaho, which was about as remote as any territory, were bottling beer from early on. Schrader and Hosp of Silver City, Idaho Territory, advertised bottled beer in 1867. Spielman and Co. of Idaho City sold bottled beer in 1868. Idaho City rival Riid and Brodbeck had bottled beer for sale two months later. Charles Kohny in Placerville, Idaho, advertised bottled beer in 1870.[35]

A frequent problem on the frontier was getting beer bottles returned to the brewer. Late in 1878, Idaho City brewer Nicholaus Haug issued a warning in the newspaper that he would seize his bottles wherever he found them.[36] He also began to use a new type of stopper on his bottles that year. The local newspaper described them as having a thick wire attached to the stopper and made in such a way that it could be reclosed any number of times without wearing out.[37] This sounds like a description of the lightning stopper patented by Charles De Quillfeldt of New York City in 1875.

Bottles found at archaeological sites are often broken and fragmentary and consequently difficult to identify. A dig at Johnny Wards Ranch in Arizona unearthed seventeen identifiable beer bottle necks and base fragments. They were brown or clear glass of varying thickness with hand-finished necks. Most were calculated to be 24-ounce size, but one may have been a 16-ounce bottle.[38]

Bottle Manufacture as a New Industry

In 1885, a scheme to build a glass plant in Salt Lake City was hailed as meeting a regional need. News reports revealed that Salt Lake City brewers could not keep the "local trade in bottle goods unless Eastern manufacturers ship their goods here so that Salt Lake's can use their bottles. There is one bottle factory west of the Mississippi and that is in California. The Salt Lake Brewery has kept a man throughout Utah and adjoining territories the past year collecting empty bottles by the car load for use here."[39] Over a decade later, beer bottles were in such demand in the Spokane, Washington, area that a representative of the New York Brewery in that area offered to pay forty cents per dozen for empty bottles.[40]

In other places during this time period, bottles were more common. In 1904, William F. Peck of Tonopah, Nevada, built a two-room house measuring 20 by 16 feet with the walls made by stacking empty beer bottles. The bottles were turned with the lip toward the inside, and they were held in place by dried adobe. Peck used light green bottles on one wall and dark green on the opposite wall for effect.[41] Pictures of his house were widely published.

During World War II materials were so scarce, especially metal for bottle caps, that quart "picnic" beer jugs became common. Often these had old-fashioned wired stoppers instead of Painter closures. They were often called "victory" bottles. In May, 1942, the War Production Board allowed breweries to collect tin cans and cut caps out of size ten and larger tin cans, provided the rest of the metal was turned in to the scrap metal drives.[42]

Modern bottles usually have the manufacturer's symbol, the item number, year of manufacture, cavity number, and the manufacturing plant on the base.[43] Various sources such as the small undated pamphlet "Bottle Language" published in Washington, D.C., by the Glass Packaging Institute, may aid in identification. The major source of this information is from Julian Harrison Toulouse's 1971 book *Bottle Makers and Their Marks* published by Thomas Nelson, Inc. In that volume, Toulouse has collected and described hundreds of marks.

The future of the beer bottle may be in its infancy. Miller Brewing in 2000 began widespread distribution of recyclable plastic beer bottles.[44] This could be the start of just one more variation in the material culture of brewing.

Bottle Caps and Closures

Beer bottles were first sealed with corks, as wine bottles had always been.[45] Bottles designed for use with corks have a tapered neck interior about an inch long. A hand-operated device squeezed the cork and inserted it into the neck. Thin wire then circled the bottle top and went over the cork to hold it in. Such devices are still sold for home wine makers. Beer's carbonation tended to eventually work unwired corks loose, causing the liquid to leak and the carbonation to release prematurely. Corks were expensive and the hand labor tedious to cork individual bottles. About 1850, a serious search for an alternative began.

In 1875, British inventor Hiram Codd invented a bottle with a glass marble confined inside the neck. The carbonated liquid inside the bottle would push the marble against a rubber washer to seal the bottle.[46] These were used in England but were never popular in America. A drawback was that children deliberately broke the necks so they could take the marbles for play.

Brewery supply houses still sold brewer's corks in 1891.[47] In one catalog, twelve corks 1-1/8 inches in diameter by 1 inch long cost $0.61. Twenty-six corks 2 inches in diameter and 2 inches in length cost $5.04. Also available was a "Steam Cork Washer" for sale.[48] The "Steam Corking Machine" in this catalog seemed to handle only one bottle at a time, and that bottle had to be set in place by hand.[49] The lightning bottle wiring machine claimed it could process 1,200 or more bottles per hour and use less wire than with hand wiring, doing the work of five men.[50]

Henry W. Putnam of New York City patented fasteners to hold corks in beer bottles in 1857 and again in 1859. A heavy wire fastener was anchored permanently around the blob at the top of the bottle. The heavy top was forced over the top of cork and held on by the counter pressure put on the cork by the contents of the bottle. These were used from 1860 to about 1875. They tended to bounce off while being transported on bumpy wagons.

Samuel A. Whitney received patent number 31,046 on January 1, 1861 for his bottle stopper. It required large threads to be put in the top of the bottle, the stopper was molded glass with a cork washer, and the top was apparently screwed into place. This was never popular with brewers.

Alfred L. Bernardin of Evansville, Indiana, invented a tin disc bottle top to allow tighter corkage. The disc, which often had grooves or tabs on it to guide the wire, went over the cork and was wired to the bottle's neck. The discs were costly and time consuming to use. Bernardin designed a wire-free disc a few years later. A metal neckband secured the Bernardin tin cap. Blob top bottles had to be trimmed or slightly redesigned so the cap fit properly. The caps were not reusable and were an added expense.[51]

The first cork-free breakthrough was the lightning stopper patented by Charles De Quillfeldt of New York City in 1875. Henry W. Putnam bought De Quillfeldt's patent and received several other patents for his improvements. The devices had a lid, bail, and a neck tie-wire. The metal lid had a rubber washer where it contacted the bottle and the top of the lid had a hole through which the bail ran. Early lids contained dangerous levels of lead, but later a patent was granted for a vulcanized, all-rubber lid.

These devices stayed attached to the bottle when the consumer opened it, and the bottle and lid were reused. The major drawback was that they were hard to clean. Sales claims boasted that "the 'Lightening' [*sic*] Stoppers are also being largely used upon beer bottles for export by parties who report that their sales of beer have been greatly increased, particularly in California and Texas . . ."[52]

William Painter, a Baltimore machine shop foreman, patented a variation in 1885 called the Triumph Lightning Stopper. These were common for about twenty-five years. Karl Hutter of New York bought the patent from Putnam in the late 1880s and continued improvements on it. On February 7, 1893, Hutter patented a revised version that used a porcelain plug for a lid. This was more sanitary and allowed advertising to be printed on the top.[53] Some of these stayed in use for decades to come. In 1885 alone, almost fifty different patents were issued for bottle stoppers.[54]

A brewer's supply catalog in 1891 sold corks and all accoutrements to go with them, lightning bottle toppers, and combined metallic cap and fastener styles. The brewing industry had not settled on a standard.

William Painter next developed the Baltimore Loop Seal —invented in 1885 and patented in 1887. A metal loop was inserted into the center of a rubber plug so that the consumer could pull it out. The plug was covered in wax cloth for cleanliness and then inserted into a ring-shaped groove inside the mouth of the bottle. It was cheap and easy to use but the bottles had to be specially molded with the internal groove. Painter supplied a special tool to glassworks to make the bottles. Painter later altered the design to have a grooved slot with a rubber band in it on an aluminum disk.

Achille Weissenthanner patented his Phoenix Cap in Paris in 1892 and showed it off at the Chicago Columbian Exhibition in 1893. It was somewhat like a crown cap and somewhat like a Bernardin metal cap. A metal band that fit into a specially shaped lip on the bottle held on the round cap. The band was inside the edge of the cap, which had a slot in it for this band. Weissenthanner supplied a hand turning tool to bottle manufacturers to make the special lip. By 1905 this system had been abandoned in America but was still widely used internationally. This cap led bottle manufacturers to develop a standardized system of bottle neck sizes still in use early in the twenty-first century.

Painter was not finished working on the problem and finally perfected a top that made everything else obsolete. The simple, economical, and sanitary Crown Cork Closure he patented on February 7, 1892, dominated the industry. He formed the Crown Cork & Seal Company. The automatic crown power machine was being advertised in catalogs in

1891.[55] The greatest drawback was that improperly formed bottles leaked. In 1903 Michael J. Owens invented the automatic bottle machine that made standardized crown bottles and solved that drawback. Beginning in 1905, the Owens machine took over the industry and by 1910, no one manufactured blob top bottles anymore.

The Perfection Bottling Machine, an example of which was used by the Klondike Brewery of the Yukon, was specifically made to "fit and fill bottles of different closure types," which was especially valuable in small brewing operations.[56]

William Painter's standard 21-flute crown or bottle cap with cork backing dominated for seventy years.[57] The short skirt crown with plastic backing was developed in the 1950s and first appeared in 1962. It captured the entire market and the old style was no longer manufactured after 1968.

A standard sheet of tinplate for pressing measured 35 inches by 28.5 inches. The original design had 21 rows, each 22 crowns wide, for 462 crowns to be punched simultaneously by the press. The short skirt type had 22 rows of 23 crowns each or 506 crowns per sheet, thus 44 more crowns. With minimum orders of 250,000 crowns, nearly 50 sheets of tinplate were saved. The "twist off" crown was developed in 1971–72. In 2006, crowns were ordered with a choice of four different plastics as liners, each of which required a different amount of torque to remove.[58]

Problems have occurred periodically with bottle caps, such as in 2001 when Coors Brewing Co. sued its cap supplier Zapata Industries. Coors claimed the caps made the beer taste like cardboard.[59] In 1999 Anheuser-Busch recalled 6 million bottles of Budweiser sold in Europe because the tops of the bottles were slightly larger than they should have been and could break when opened.[60]

Beer Can Openers

Once the beer was in the cans, consumers needed a way to get it out. In 1932 American Can Company perfected a device for opening cans of liquid that they named the "Quick & Easy" opener. Openers were developed simultaneously with cans until the advent of self-opening cans. Openers were made in many styles, shapes, sizes, and with brewery and/or beer names on them. Bottle opening devices or cap lifters had existed for decades. Before long, devices with a cap lifter at one end and a punch device at the other was developed to accommodate both bottles and cans.

Beer Cans

The first recorded use of tin cans to preserve food in the United States was documented in 1839.[61] Canners put milk in cans early on, but beer had to wait.

Development of the Beer Can

Cans for beer were experimented with as early as 1909 when a Montana brewer asked if the concept was feasible, but they proved unsuccessful because there was no lining capable of protecting the taste of the beer from contamination by the metal. Before Prohibition was over, Anheuser-Busch, Pabst, and Schlitz experimented with cans. Erwin Uihlein, president of Schlitz, "designed his own version of a beer can that looked like a stout little barrel."[62] The prohibitive cost put an end to these experiments. Durability was another major consideration; cans designed to hold vegetables needed to be sturdy enough to withstand 25 to 35 pounds of pressure per square inch; beer could exert 80 p.s.i.[63]

Can companies worked on developing a can coating to keep the beer from contacting the metal, but decided that no single coating could meet all the requirements. The coating had to chemically resist the beer, tolerate pasteurization temperatures of 140 degrees, and endure refrigeration of less than 40 degrees. For a while, the American Can Company planned to enamel the cans and then spray them with brewers pitch, just as kegs had been treated for centuries.[64] Then, Union Carbide and Carbon Corporation developed a plastic coating for the American Can Company that they named Vinylite. American Can used this material in prototype cans in 1933. It adopted the coating and trademarked its beer can system as "Keglined" on September 25, 1934.[65]

Cans were marketed to brewers as providing many benefits, such as keeping light out and eliminating its harmful effect on the beer, and being easily cooled, compact, nonbreakable, and disposable. In late 1933, the Gottfried Krueger Brewing Company of Newark, New Jersey, agreed to try selling canned beer. On January 24, 1935, the first cans of its new canning line rolled out, destined for a test marketing campaign in Richmond, Virginia. Richmond was on the fringe of the marketing area and had low per capita beer consumption.[66] Canned beer was a hit; Krueger was filling 180,000 cans a day by August that year.

Pabst was the first of the big three companies to begin canning beer. By mid-1935 National Can Company had entered the competition with a can that was double-coated with enamel. In 1936, in addition to its cap-sealed, cone-top cans, Continental Can Company began to sell flat-top cans with a lining system that used hard wax over an initial coating of enamel.

To reassure cautious beer drinkers, can labels stated: "the beer does not come in contact with metal. A scientifically safe lining keeps the beer 'barrel fresh.'" About one can in five thousand suffered excess porosity in the lining and the beer turned bad. This was an acceptable rate of failure.[67]

The Stubby

Beer bottle manufacturers did not stand around and accept the new competition. Owens-Illinois, the largest bottle maker, designed a short, stocky, lightweight, non-returnable bottle they named the Stubby.[68] By the end of 1935, twenty-eight more brewers were using them than were experimenting with cans.

Variations in Can Production

In 1935 Pacific Can Company began producing cans for west coast brewers.[69] Each can manufacturer had patents for the unique features of its design. Continental's cone top used a conventional bottle cap and could be filled on bottle lines. Many small brewers could not afford canning lines, and this situation remains to this day.

Early canning lines filled 160 cans per minute compared to 120 bottles. Over an eight-hour shift, the difference was great. A beer delivery truck of the 1930s could carry 400 cases of cans but only 200 cases of bottles; this difference helped brewers absorb the greater production cost of cans over bottles.[70] In the mid 1930s, as a general rule, locally brewed beers were sold at retail for ten cents per bottle or three for twenty-five cents. Canned beer at three for thirty cents was a bit higher. National brands sold for a "premium" at fifteen cents each.[71]

Can dating may be relatively simple—some brewers put the date of production on the can—or quite convoluted. Some of the factors that reveal a can's date are the style of the top, the form of the bottom, the design of the can, the size of the can opener displayed on the side, the design of the can seam, the presence of a two-step or four-step opening instruction panel, the Internal Revenue Tax Paid Statement, a Withdrawn Free of Internal Revenue Tax statement, and the type of paint used on the can. Some states required the state permit numbers on the cans from 1935 to 1937, and some states had the lid stamped with the State Beverage Tax Seal.[72]

Markings on Beer Cans

American Can Company of Greenwich, Connecticut, put "American Can Company," "A.C.CO.," "AM. CAN CO.," or similar lettering on its cans from 1933 to the mid-1950s. This company's cans read "Patent Pending" from early to late 1935. In late 1935 the cans were marked "No Patents." From early 1936 to mid-1937, cans read "Patents Pending." From mid to late 1937, the imprint was "Patents 1625229–2064537 others pending." Next came "Pats. 1625229–2064537 others pending" from 1938 to 1947.[73] The company also used a Manufacturers Printing Code (MPC) on each can. The number in the code represented the specific plant at which the can was manufactured—fifteen plants were operated around the country.

Continental Can Company of New York City used "C.C.CO." and "PAT.APPL'D FOR" on its cans from 1936 to 1945.[74] Crown Cork & Seal Company of Philadelphia printed "CROWN CAN CO." on its cans from 1937 to 1938.[75] National Can Company of Chicago put "NATIONAL CAN CO." or "N.C.C." on cans from 1935 to 1938. It used the underlined term "<u>double</u> lining" or "<u>double lining</u>" from mid to late 1935. "<u>DOUBLE-LINED</u>" and "Patents Pending" appeared from December 1935 to 1938.[76]

National Can Corporation of New York City was the renamed and relocated National Can Company. After the move, the cans had "NATIONAL CAN CORP." printed on them from 1937 to 1945. The term "<u>DOUBLE-LINED</u>" was printed in a small box along with patent and corporation information and "Patents Pending."[77]

Pacific Can Company of San Francisco printed "PACIFIC CAN CO. S.F. CAL." on its cans from 1935 to 1948. In 1949 it printed "PACIFIC CAN CO." It used the trademark "KEG-LET" from 1938 to 1948. In 1949 the new trademark was "Kan-Kegs."[78] This company's cans were initially lidded with soup can lids until a better lid, capable of withstanding beer pressure, was developed. The soup lids had a bead curving upward that encircled the whole lid and was used on both the tops and bottoms of the cans. By late 1935 a standard flat lid was introduced.

Can manufacturers used both metallic and enamel paints prior to World War II. Congress restricted the use of various metals during the war, and most canning of beer stopped. Dull gray and olive drab cans were produced for beer shipped to soldiers overseas. After the war, metallic paints were again available and many brewers used them to celebrate with very bright cans from late 1945 into early 1946.

Evolution of Beer Cans

Can lids that were perfectly flat were on all beer cans from 1935 to 1937. In 1937, top lids with a bead around them—somewhat like the soup can lids that had been briefly used—were found to withstand pressure better, and became the norm. Can seams were straight, single grooved, crimped, and then soldered. This design was used on the early cans and into the 1950s. American Can Company developed a crimped seam sealed with epoxylike glue. These have a square-toothed look. They first appeared in 1938 on 32-ounce cans and later in World War II on 12-ounce cans. Only a few brewing companies used these.

An example of the potential dating of beer cans was offered in a regular column in a breweriana collector's club magazine in 2006.[79] Five unusual specimens were dated to within a range of a few years or down to the exact hour of production. A Ballentine Export Beer produced in Newark, NJ, in a 32-ounce cone top can was found with a series of

Ohio tax stamps still affixed. The can dated between the late 1930s and early 1940s. A Narragansett Lager Beer can from Cranston, RI, bore a paper label that read "Flip Top," apparently because the cans painted with those words had not yet arrived at the brewery. It was dated to 1962. An Eastside Old Tap can produced by Pabst Brewing, Los Angeles, had a sticker attached that stated "The Alcoholic Beverage contained herein is distressed merchandise, salvaged from fire, flood, wreck or similar catastrophe." It dated to the late 1960s to early 1970s. A Hamm's Beer can shipped from St. Paul, MN, was one of the first all-aluminum cans successfully filled on the new canning line on February 13, 1964. A Griesedieck Light Lager Beer can from St. Louis, MO, was stamped with a date and number on the top. It was can number 100,000,034 to be filled and the line was stopped to celebrate the 100,000,000th can at 4:25 P.M. on December 27, 1950. A number of cans were dated and given as souvenirs.

Charles Sollberg of New York received patent 1,625,229 on April 19, 1927, for his new method of making the seams on the sides of cans in a way that allowed air to escape during soldering.[80] Walter E. Groenke of Maywood, Illinois, received patent 2,064,537 on December 15, 1936, for a way to reinforce the side seams of cans to withstand internal pressure such as beer would produce. William E. Taylor from East Orange, New Jersey, received patent 2,178,618 on November 7, 1939, for his invention to add a protective lining strip to the inner side seam of cans. Russell C. Taylor from Greenwich, Connecticut, received patent 2,259,498 for an improved side seam made with alternate interlocking tabs and able to withstand high pressure.

The Crowntainer type of cone top was filed for patent on May 13, 1940, and was approved on September 18, 1945, with patent number 2,384,810.[81] The application had nine pages of text, four pages of detailed drawings, and made thirty-five claims pertaining to the uniqueness of the design.

Collector Jack Martells, after years of study, found that cone tops could be dated to four different periods: Prewar from 1935 to 1941; the War Years from 1942 to 1945; Postwar-IRTP from 1945 to 1950; and Non-IRTP from 1950 to 1956.[82] Prewar cans included Continental Can's low spout, Crown Cork's three-piece high neck, and the Crowntainer. War Years cans were, with few exceptions, the olive-drab cans with almost unreadable brand names, with "Withdrawn Free From Internal Revenue Tax for Exportation" printed on them. Postwar cans are marked I.R.T.P., meaning "Internal Revenue tax paid." These include high spout Continental, three-piece Crown cans, and Crowntainers with patent numbers. Non-IRTP cans reflect the changed government rules that no longer required tax information on each can after March 1, 1950. Copyright dates printed on the can reflect the date that the brand name was copyrighted, and not the date of the manufacture or filling of the can.

A researcher can date unlabeled cone tops by measuring the rim diameter, height to rim, overall height, weight, and design. Martells's book charted this information.[83]

Beer cans changed and developed in minor ways, and continue to do so as of the time of this writing. In the 1950s, for example, the cone top can began to lose ground because it could not be filled at the brewery as quickly as flat top cans. Retailers had difficulty stacking and displaying cone tops. Meanwhile, consumers had learned to open flat top beer cans without reading the directions first. During this decade, cans of various sizes appeared. Schlitz introduced 16-ounce flat tops in 1954, while several others introduced 8-ounce cans.[84]

In the 1960s, tab tops and aluminum were the innovations in beer cans. Aluminum tops for cans made a lighter can and saved transportation costs. Alcoa Aluminum patented the pull tab in 1962. The device was variously known as the zip top, pull tab, tab top, snap top, and ring top. In 1962, Pittsburgh Brewing's Iron City Beer test marketed the tab top in Virginia.[85] In 1963, Schlitz began to use these and advertise them widely. Jack Martells recorded seven major and eighty minor changes in tab tops up to 1976; his book illustrates most of these.[86]

In 1959, Coors marketed an all-aluminum can, and Primo of Hawaii made aluminum cans with paper labels that proved to be a disaster. In 1963, 12-ounce aluminum cans were test marketed for Hamm's and Anheuser-Busch. By 1967 aluminum cans accounted for more than 10 percent of the cans sold.

Manufacturers of steel cans rose to the competition. An ongoing problem was the unpredictable fluctuation in the price of tin. The industry therefore developed tin-free steel (TFS), which used a chemical coating. To make cans of the new material, an adhesive seam construction called Mira-Seam was developed in 1965 and was generally available the next year.[87] To compete, Continental began making a solderless can with a welded seam it dubbed Conoweld.

Tab tops littered America in a way that became intolerable, and a push for an alternative began. Push-button tops developed, in which a small button was pushed to release the pressure and a large one was opened to become the drinking hole. Other tabs were developed that stayed with the can rather than being detached when opened. These were termed "stay behind" technology and were intended to reduce litter. Tab tops will probably be found in sites that date from the later twentieth century for the foreseeable future.

In the 1980s, cans with UPI computer codes first appeared. In 1989, government warnings about the alleged harmful effects of alcoholic beverages were included on the labels.[88]

A new packaging system for beer, which did not reach American breweries until the twenty-first century, was the "aluminum beer bottle."[89] This is a creature of mixed ancestry, part bottle and part can. It is all aluminum, shaped like a beer bottle, and closed with a bottle cap. Its appeal is that the beer stayed cold for up to an hour longer than conventional packaging. [90]

As D. B. S. Maxwell has written, "Changes in beer can design are sufficiently well documented to render this artifact very useful as a dating tool. Its presence on virtually all parts of the landscape also suggests that it should be employed by archaeologists interested in inferring the age of historic deposits, or the date of intrusion on prehistoric sites."[91] In 1969, an estimated "1,809 million" cans, mostly beverage cans, littered America's roads.[92] As Jane Bush said, "The historical archaeologist who ignores the beer can at his site is like the prehistoric archaeologist who ignores historic pottery."[93]

NOTES

1. Bob Kay, "Evolution of the Budweiser Label," pamphlet published by the author, January 1993.
2. Ed Love and J. Peterson Powell, *Brewery and Beer Trademarks, vol. I* (Colorado Springs: Villa Publishing, 1989). Ed Love and J. Peterson Powell, *Brewery and Beer Trademarks, vol. II* (Colorado Springs: Villa Publishing, 1991).
3. Kevin C. Lilek, *United States Beer Cans with Opening Instructions, The Standard Reference Guide of Opening Instruction Flat Tops* (Aurora, Illinois: Lilek Publishing, 1998). Kevin C. Lilek, *United States Beer Cans with Opening Instructions, The Standard Reference Guide of Opening Instruction Flat Tops (Supplement)* (Aurora, Illinois: Lilek Publishing, 2006).
4. Millard, *History of the Carson Brewing Company,* 69.
5. Hennech and Pierson, *Beer Trucks, A Photographic Salute, 1870–1992,* 64, 65.
6. Hennech and Pierson, *Beer Trucks, A Photographic Salute, 1870–1992,* 64, 65.
7. *Lewiston Morning Tribune* (Lewiston, Idaho), August 28, 2002, Section B, p. 1, c.1.
8. *Real Beer Page Mail, The Free Monthly Beer News Digest for the Online Beer Enthusiast,* March, 2002, p. 4.
9. Michael Polak, *Bottles: Identification and Price Guide,* 2nd ed. (New York: Avon Books, 1997), 65.
10. Polak, *Bottles: Identification and Price Guide,* 7.
11. "Latter-Day Shakers," *Archaeology* (January/February 1999): 28–29.
12. Guetig and Selle, *Louisville Breweries: A History of the Brewing Industry in Louisville, Kentucky, New Albany and Jeffersonville, Indiana,* 14.
13. Wayne L. Kroll, *Badger Breweries Past and Present* (Jefferson, WI: Wayne L. Kroll, 1976), 1.
14. Ron Feldhaus, *The Bottles Breweriana and Advertising Jugs of Minnesota, 1850–1920* (Minneapolis: North Star Historical Bottle Collectors Association, 1986), 9.
15. Kroll, *Badger Breweries Past and Present,* 3–4.
16. Apps, *Breweries of Wisconsin,* 82.
17. Guetig and Selle, *Louisville Breweries: A History of the Brewing Industry in Louisville, Kentucky, New Albany and Jeffersonville, Indiana,* 39.
18. Byron Martin and Vicky Martin, *Here's to Beers: Blob Top Beer Bottles, 1880–1910* (Northridge, California: The Achin Back Saloon, 1973), 11.
19. Plavchan, "A History of Anheuser-Busch, 1852–1933," 75.
20. Plavchan, "A History of Anheuser-Busch, 1852–1933," 76.
21. Plavchan, "A History of Anheuser-Busch, 1852–1933," 76.
22. Holian, *Over the Barrel: The Brewing History and Beer Culture of Cincinnati, 1800 to the Present. Volume One (1800–1919),* 130.
23. Polak, *Bottles: Identification and Price Guide,* 2nd ed., 75.
24. Michael Polak, *Bottles: Identification and Price Guide,* 4th ed. (Iola, Wisconsin: Krause, 2002), 16.
25. Kroll, *Badger Breweries Past and Present,* 6.
26. Polak, *Bottles: Identification and Price Guide,* 4th ed., 18.
27. Burley and Will, "Special Brew: Industrial Archaeology and History of the Klondike Brewery," 77.
28. Guetig and Selle, *Louisville Breweries: A History of the Brewing Industry in Louisville, Kentucky, New Albany and Jeffersonville, Indiana,* 39.
29. Polak, *Bottles: Identification and Price Guide,* 4th ed., 21.
30. Polak, *Bottles: Identification and Price Guide,* 4th ed., 21.
31. Maxwell and Sullivan, *Hometown Beer: A History of Kansas City's Breweries,* 114.
32. Paul Burden, *The Guide to U.S. Painted Label Beer Bottles* (Dover, MA: Norfield Publishing, 1990), p. 3.
33. Burley and Will, "Special Brew: Industrial Archaeology and History of the Klondike Brewery," 74.
34. Burley and Will, "Special Brew: Industrial Archaeology and History of the Klondike Brewery," 84.
35. *Owyhee Avalanche* (Silver City, Idaho), March 30, 1867. *Idaho World* (Idaho City, Idaho), May 6, 1868 and July 1, 1868.
36. *Idaho World* (Idaho City, Idaho), December 17, 1878, p. 3, c. 2.
37. *Idaho World* (Idaho City, Idaho), December 17, 1878, p. 3, c. 1.
38. Bernard L. Fontana and J. Cameron Greenleaf, et al., "Johnny Ward's Ranch: A Study in Historical Archaeology," *The Kiva* 28:1–2.
39. *Idaho Register* (Idaho Falls), Aug 29, 1885. The report was from a Salt Lake City newspaper.
40. New York Brewery, Spokane, Spokane Breweries Records, Washington State University, Cage 7, Box 3, Letter Book 1, Oct. 1, 1895 to June, 1896, p. 453.

41. *Idaho Daily Statesman* (Boise, Idaho), January 21, 1904, p. 3, c. 2–4.
42. Apps, *Breweries of Wisconsin*, 151.
43. "Bottle Language," pamphlet (Washington, D.C.: Glass Packaging Institute, n.d.), 2.
44. *Real Beer Page Mail, The Free Monthly Beer News Digest for the Online Beer Enthusiast*, rbpmail@realbeer.com, April 2000.
45. Morris and Johnson, *Richmond Beers, A Directory of the Breweries and Bottles of Richmond, Virginia*, 2nd ed., 100–103. The discussion of closures is all from this source unless otherwise noted.
46. Polak, *Bottles: Identification and Price Guide*, 4th ed., 19.
47. Morris and Johnson, *Richmond Beers, A Directory of the Breweries and Bottles of Richmond, Virginia*, 2nd ed., 210.
48. Morris and Johnson, *Richmond Beers, A Directory of the Breweries and Bottles of Richmond, Virginia*, 2nd ed., 222.
49. Morris and Johnson, *Richmond Beers, A Directory of the Breweries and Bottles of Richmond, Virginia*, 2nd ed., 224.
50. Morris and Johnson, *Richmond Beers, A Directory of the Breweries and Bottles of Richmond, Virginia*, 2nd ed., 225.
51. An advertising page from the Bernardin Bottle Cap Co., Evansville, Ind., showed the Bernardin cap and the Cyclone Red Velvet Rubber stoppers with eight different stopper tops from which to select. See Morris and Johnson, *Richmond Beers, A Directory of the Breweries and Bottles of Richmond, Virginia*, 2nd ed., 135.
52. Morris and Johnson, *Richmond Beers, A Directory of the Breweries and Bottles of Richmond, Virginia*, 2nd ed., 227.
53. Blue Seal Supply Co. of Portland, Mass., sold these for $1.25 per gross. They also carried Blue Seal cork fasteners and several others. See Morris and Johnson, *Richmond Beers, A Directory of the Breweries and Bottles of Richmond, Virginia*, 2nd ed., 134.
54. A drawing showing these from the 1886 *Bottles Gazette* was reproduced in Hull-Walski and Walski, "Brewing and Bottling in Harpers Ferry, West Virginia," p. 17.6.
55. Morris and Johnson, *Richmond Beers, A Directory of the Breweries and Bottles of Richmond, Virginia*, 2nd ed., 230.
56. David V. Burley and Michael H. Will, " 'The Beer That Made Klondike Famous and Milwaukee Jealous': The O'Brien Brewing and Malting Company Site, Klondike City, Yukon," *Industrial Archaeology* 26, no. 1 (2000): 50.
57. Painter applied for patent on May 19, 1891, for a Bottle-Sealing Device. Patent 468,226 was granted.
58. John Vetter, "What is Unique About These Crowns?" *American Breweriana Journal* 144 (November-December 2006): 42.
59. *Real Beer Page Mail, The Free Monthly Beer News Digest for the Online Beer Enthusiast*, rbpmail@realbeer.com, September 2001.
60. *Real Beer Page Mail, The Free Monthly Beer News Digest for the Online Beer Enthusiast*, rbpmail@realbeer.com, September 1999.
61. Jane Busch, "An Introduction to the Tin Can," *Historical Archaeology* 15, no. 1:96.
62. Jack Martells, *The Beer Can Collector's Bible* (New York: Ballentine Books,1976), 6.
63. Bruce Gregg, John Page, and Tom Waggoner, eds., *United States Beer Cans: The Standard Reference of Flat Tops and Cone Tops* (Fenton, MO: Beer Can Collectors of America, 2001), 7.
64. Gregg, Page, and Waggoner, eds., *United States Beer Cans: The Standard Reference of Flat Tops and Cone Tops*, 8.
65. Kevin C. Lilek, *United States Beer Cans with Opening Instructions, The Standard Reference Guide of Opening Instruction Flat Tops* (Aurora, Illinois: Lilek Publishing, 1998), iv.
66. Martells, *The Beer Can Collector's Bible*, 7.
67. Martells, *The Beer Can Collector's Bible*, 8.
68. Martells, *The Beer Can Collector's Bible*, 8. 69. Lilek, *United States Beer Cans with Opening Instructions, The Standard Reference Guide of Opening Instruction Flat Tops*, iv, v.
70. Martells, *The Beer Can Collector's Bible*, 8.
71. Martells, *The Beer Can Collector's Bible*, 8.
72. Lilek, *United States Beer Cans with Opening Instructions, The Standard Reference Guide of Opening Instruction Flat Tops*, v.
73. Lilek, *United States Beer Cans with Opening Instructions, The Standard Reference Guide of Opening Instruction Flat Tops*, vii, viii.
74. Lilek, *United States Beer Cans with Opening Instructions, The Standard Reference Guide of Opening Instruction Flat Tops*, vii.
75. Lilek, *United States Beer Cans with Opening Instructions, The Standard Reference Guide of Opening Instruction Flat Tops*, vii.
76. Lilek, *United States Beer Cans with Opening Instructions, The Standard Reference Guide of Opening Instruction Flat Tops*, ix.
77. Lilek, *United States Beer Cans with Opening Instructions, The Standard Reference Guide of Opening Instruction Flat Tops*, ix.
78. Lilek, *United States Beer Cans with Opening Instructions, The Standard Reference Guide of Opening Instruction Flat Tops*, ix.
79. Dave Babel, "Beer Can Closeup," *Beer Cans and Brewery Collectibles* (December 2006/January 2007): 36–37.
80. Kevin C. Lilek, *United States Beer Cans with Opening Instructions, The Standard Reference Guide of Opening Instruction Flat Tops, Supplement* (Aurora, Illinois: Lilek Publishing, 2006), v.
81. Jack Martells, *The Cone Top Collector's Bible* (Matteson, Illinois: Great Lakes Living Press, 1976), 8.
82. Martells, *The Cone Top Collector's Bible*, 16.
83. Martells, *The Cone Top Collector's Bible*, 19.

84. Martells, *The Beer Can Collector's Bible,* 9.
85. Martells, *The Beer Can Collector's Bible,* 10.
86. Martells, *The Beer Can Collector's Bible,* 14–21.
87. Martells, *The Beer Can Collector's Bible,* 10.
88. D. B. S. Maxwell, "Beer Cans: A Guide for the Archaeologist," *Historical Archaeology* 27, no. 1 (1993): 111.
89. These were marketed in Europe in the 1990s, and American collectors acquired some as souvenirs.
90. Charles Sheehan, "It's the Bottle . . . and a Lot More," *Lewiston Morning Tribune* (Lewiston, Idaho), August 25, 2004, 12B.
91. Maxwell, "Beer Cans: A Guide for the Archaeologist," 111.
92. Busch, "An Introduction to the Tin Can," 100.
93. Busch, "An Introduction to the Tin Can," 100.

APPENDIX I

A Chronology of Brewing in the New World

1502 Christopher Columbus noted the presence of fermented beverages brewed by Native Americans.

1544 Juan Herrera wrote to the king of Spain to tell of establishing a brewery in Mexico City—the first European brewery in the New World.[1]

1587 English colonists in Roanoke, Virginia, brewed ale from corn (maize).[2]

1607 First shipment of beer from England reached Jamestown, Virginia.[3]

1609 Advertisements appeared in London asking for brewers to move to Virginia.

1613 Dutch explorer Adrian Block erected several houses on Manhattan Island, one of which became a brewery. Some sources refer to the brewery of Block and Hans Christiansen as being built in 1612.

1617 Louis Herbert emigrated to Quebec and grew barley. He and his wife brewed beer for friends, neighbors, and ceremonial purposes. He was Canada's first brewer.[4]

1620 Pilgrims disembarked in Plymouth because the beer supply on the ship was low and they dared not sail farther.

1629 The Massachusetts Bay Company ordered hop seeds from England.[5]

1632 The Dutch West Indian Company erected a brewery in New Amsterdam to brew beer from oats.

1633 Kilian Van Rensselaer established a brewery in Albany.

Peter Minuit established a brewery at Market Field on Manhattan Island.[6]

Governor Van Twiller of New Netherlands established a brewery near the future Wall Street.[7]

1637 First well-documented brewery was established in the Massachusetts Bay Colony under the control of Captain Sedgwick.[8]

1638 Governor Kieft of New Netherlands forbade tapsters from retailing wines and liquors. Since only beer could be sold, brewing expanded greatly.[9]

1639 Sergeant Baulston was placed in charge of the brew house in Providence, Rhode Island.[10]

Harvard students revolted over the insufficiency of beer provided with their meals.[11]

1646 The prices of domestic and imported hops were regularly posted at markets in Manhattan.[12]

Jesuits constructed a brewery at Sillery, near Quebec, Canada.[13]

1652 In Virginia, George Fletcher obtained exclusive rights to "brew in wooden vessels, which none had experience in but himself."[14]

1654 Peter Lindstrom, an engineer of New Sweden, said locals brewed strong clear beer from maize, and beer or brandy from persimmons.[15]

1657 A brewery owned by F. B. Pastoor was sold in New York City with house and lot for 3,600 Guilders. The outfit consisted of an "undertrough, a cooler, a brew kettle, a wort tub, a barrel kettle and twenty half barrels and beer tubs, an iron vise, a little kettle, a wooden funnel, a brewer's sling, two gutters, a long and a short gutter, and a scoop."[16]

1662 John Winthrop wrote to the Royal Society of London explaining that malt made of corn was delicate and the sprouts had to grow very long to be effective.[17]

Figure AI.1. William Penn's brew house. (Salem 1880 [1972], 132–133.)

1664 The Duke of York, in control of former Dutch holdings, limited brewing to those with sufficient knowledge of the craft.

1667 Massachusetts passed laws stating that beer must be made of malt and not diluted with sugar or molasses.[18]

1670 Samuel Wentworth of Portsmouth obtained the first license to brew beer in New Hampshire.[19]

1683 William Penn erected a brewhouse at his estate of Pennsbury.

William Frampton erected the first brewery in Philadelphia on Front Street.[20]

1685 Penn wrote that mault [sic] was being made, and mault drinks were becoming common in his area of Pennsylvania (fig. A1.1).

1693 William and Mary College in Virginia had a rule that "No liquors shall be furnished or used at the table except beer, cider, toddy, or spirits and water, and these only in moderate quantity."[21]

1698 A congregation of Baptists began to meet at the brewery of Anthony Morris in Philadelphia, and continued to use the brewery as their home until 1707.[22]

1734 Mary Lisle, the first brewster (female brewer) in America, took over her late father's Edinburgh Brewhouse in Philadelphia, which she operated until 1751.[23]

1735 Governor James Oglethorpe banned hard liquor and encouraged settlers in Georgia to drink English beer. He offered each new settler forty-four gallons.[24]

1738 First brewery in the deep South was built at Jekyll Island, Georgia, founded by Major William Horton.

1739 First successful American glasshouse started by Casper Wistar in New Jersey.[25]

1744 Globe Brewery in Baltimore was built as a one-story structure atop three remarkable brick vaults measuring 120 by 75 by 9 feet that remained in use at least 150 years.[26]

1747	Hops were a prominent import in New York.[27]
1749	Moravian brewery built at Christian Springs, PA.[28]
1757	George Washington entered a beer recipe in his notebook.[29]
1765	The British Army built a brewery at Fort Pitt, the first British brewery west of the Allegheny Mountains.[30]
	First brewery outside the original colonies established at Kaskaskia, Illinois, by Mr. Beauvis [Beauvais].[31] This was a French town east of the Mississippi River at that time.
1772	Porter style beer was first brewed in England but exports to America failed to be popular.[32]
1774	Boston Port Act stopped imports into Boston, and thus encouraged local breweries to thrive in response.[33]
1775	General Washington recommended his troops be given one quart of spruce beer per day.[34]
1776	Brewer and tapster Israel Putnam of Brooklyn, Connecticut, became a general in the revolution.[35]
1778	British explorer Captain Cook brewed beer on shore at Vancouver Island (now part of Canada) from malt he had carried on board. He mistakenly believed that malt prevented scurvy.
	Virginia Governor Patrick Henry served home-brewed beer to important guests during the Revolution.[36]
1788	Dr. Benjamin Rush termed beer and cider Federal beverages and distilled spirits anti-federal.[37]
1789	John Nancarrow started a malting business and brewery at Scott's Landing, Kentucky, the first in the state.[38]
	The Massachusetts legislature exempted brew houses from taxes for five years.
1791	The Excise Act of 1791 taxed whiskey but not malt beverages.[39]
1797	Traveler Louis Phillipe recorded in his "Diary of My Travels in America" that he looked forward to staying in German homes where "beer is available . . . and theirs are the only inns where we have been able to buy it."[40]
	Screw presses to bale hops were introduced into Massachusetts from England about this year.[41]
1798	Dr. Richard Shannon patented a device to surround mashing tubs with steam to heat them.[42]
1800	By this time, a device had been invented to cool wort by running it through pipes immersed in cold water after preliminary cooling on the surface cooler.[43]
	About this time, W. M. Ellis of New Jersey patented a mashing machine and installed one in the brewery of J. W. Morris in Philadelphia.[44]
	Brewery architecture showed simplicity and a lack of all scientific construction. Breweries had the same character as their surroundings.[45] In older cities, some breweries were of superior construction.
1801	I. E. Boardley of Philadelphia described a home brewing vessel with three levels that he had invented.[46]
1805	J. Collins of the Lewis and Clark Expedition brewed beer from molded and soured camas root bread. Captain Clark pronounced it very good.[47]
1806	Massachusetts mandated compulsory inspection and grading of the hops grown there, which impressed European buyers.[48]
1810	One hundred forty breweries existed in the United States, and production reached 183,000 barrels. In contrast, one large brewery in London brewed 235,000 barrels.[49]
	Jacques Delassas de St. Vrain opened the first brewery in St. Louis, Missouri.[50]
	Benno Scharl, monk and brewer of Munich, Germany, separated out and reused yeast that settled to the bottom of his fermentation vessel and began brewing the beer that came to be called lager beer.[51]
1812	Davis Embree finished construction of the first commercial brewery in Cincinnati.[52]
1813	The Domestic Encyclopedia published in Philadelphia illustrated the layout of a contemporary brewery (fig. A1.2).
	British soldiers most likely burned Joseph Webb's brewery when they burned the rest of Black Rock near Buffalo, New York.[53]
1817	Harmonists open a brewery in New Harmony, Indiana.
1820	Frances Perot used a steam engine in his brewery in Philadelphia. His first one lasted fifty-three years and was a local curiosity.[54]
	The St. Louis Brewery expanded to include granaries to hold barley contracted from farmers.[55]
	Zoar Society in Ohio opened a brewery.

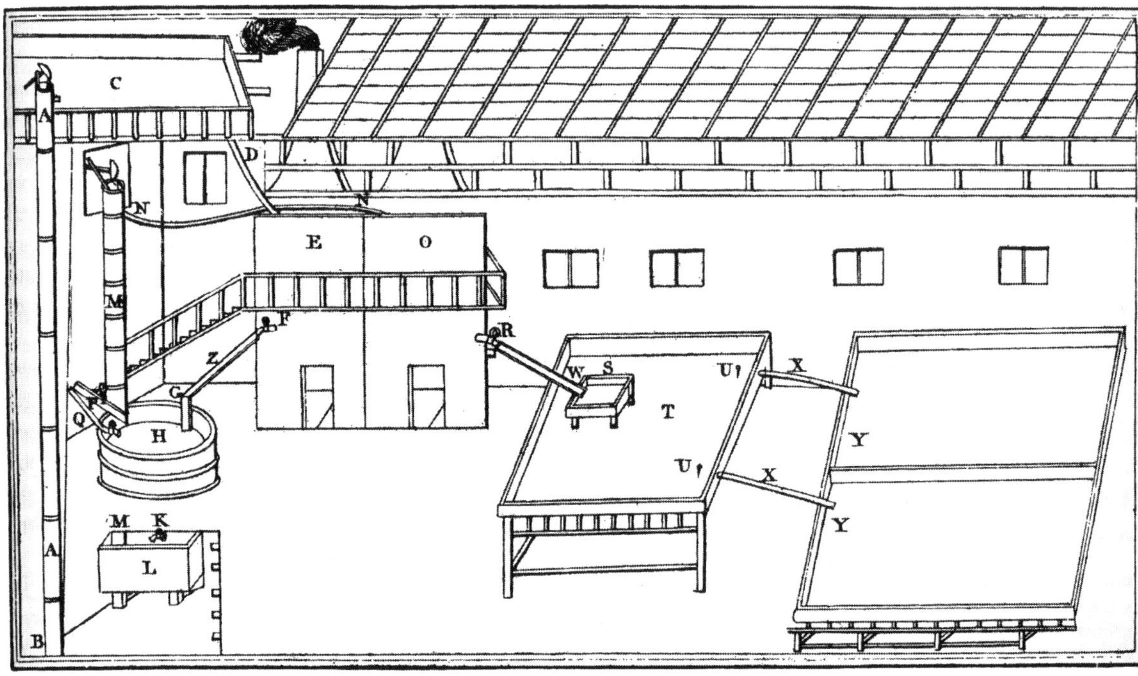

Figure AI.2. 1813 brew house with all equipment. (100 Years of Brewing, 1903, 69.)

1821 Edmund McManus bottled porter from his St. Louis Brewery at his porter cellar on Main Street.[56]

1823 Jacob Salmon at Madison, Indiana, began one of the first breweries in the "West."[57] It was built of stone and logs on a steep bank and appeared to be a single story in front but a two-story in back, and served as a residence as well as a brewery[58] (fig. A1.3).

Mullanphy's brewery in St. Louis was built of stone.[59]

1824 Fiddler & Taylor Brewery of Albany, New York, produced 250 barrels of beer per day to lead the nation.[60]

1828 In John Beveridge's new brewery in Newburg, New York, the mash tub had a mashing machine, consisting of rakes powered by steam.[61] This is one of the first uses of steam in an American brewery.

1829 David G. Yuengling founded a brewery in Pottsville, Pennsylvania, that became the longest continuously operating brewery in the United States (fig. A1.4).

Twenty-five barrels of beer from Detroit were shipped to Cleveland.[62]

1832 First well-documented brewery in Cleveland was founded.[63]

1833 First brewery in Chicago was started.

1835 Dr. Ives of New York City isolated the resinous globules from hops that he called lupulin.[64]

John Phillips started Wisconsin's first brewery at Mineral Point.[65]

1836 Matthew Vassar built a new, expanded brewery for the second time in Poughkeepsie.[66] Eventually his plant covered 35,000 square feet.

1837 Rice and Kroener established the first brewery in Evansville, Indiana.[67]

William McGlove, known as "Billy the Brewer," was credited with being the first brewer in California.[68]

National economic panic this year and subsequent depression until 1843 retarded the growth of the brewing industry in Cincinnati.[69]

1838 Frederick Schaefer arrived in New York. In 1842 he founded a major brewery with his brother.

Some evidence suggests Adam Lemp brewed lager beer in St. Louis this year.

Figure AI.3. Drawing of Salmon Brewery, Madison, Indiana. (*100 Years of Brewing*, 1903, 196.)

Figure AI.4. The Yuengling Brewing Company was the oldest still in operation in the United States to date. (*100 Years of Brewing*, 1903, 197.)

Figure AI.5. F & M Schaefer Brewery in 1842. (F & M Schaefer Brewing Co., To Commemorate Our 100th Year, 1842–1942 (New York: n.p., 1942).

1840 John Wagner brewed lager style beer in Philadelphia, often called the first example of the style in America. Lagers require cooler cellars for fermentation than ale.

Cincinnati had eight breweries.

By this decade, breweries in St. Louis and Cincinnati were often brick, never taller than three stories, and involved several structures of varying size and form.[70]

Per capita consumption of beer was 1.3 gallons per year.[71]

Hillsborough County, New Hampshire, produced more hops than any other county in America.[72]

1842 McHose and English became the first brewers to develop a natural cave in St. Louis for beer storage and as a retail outlet[73] (fig. A1.5).

1843 Stead Company of England first developed and patented a device to blow hot air through malt to clean it before steeping or kilning.[74]

First machine patented in England to clean brewers' casks.[75]

1844 Forerunners of Pabst and Miller brewing started in Milwaukee.

Mr. Balling, professor of chemistry at Prague, described a method to control fermentation scientifically by use of a saccharometer.[76]

Fortmann and Co. introduced lager beer to Cincinnati.

Sacramento's first brewery opened.[77]

1845 Adam Lemp, a St. Louis brewer, converted a natural cave along the road to Carondelet for lager cellars[78] (figs. A1.6, A1.7).

1848 Liberal revolutions in Germanic area of Europe failed, causing many Germans to emigrate to America. They brought a thirst for lager beer and experience as brewers.

1849 Adam Schuppert Brewery at Stockton and Jackson streets in San Francisco opened, which is sometimes credited (falsely) as being California's first brewery.[79]

Figure AI.6. J. Adam Lemp's brewery in St. Louis. (*100 Years of Brewing*, 1903, 212.)

Figure AI.7. J. Adam Lemp's cellar house. (100 Years of Brewing, 1903, 212.)

Steam beer was invented in California as lager brewers lacked ice and brewed at higher than normal temperatures. The unique foamy product was brewed throughout the far West for decades.[80]

Anthony Yoerg brewery in St. Paul sells the first in-state brewed lager in Minnesota.[81]

Eight pounds of hops were grown commercially in Oregon—the first in the Northwest.[82]

1850 The United States was home to 431 active breweries.

Mathias Frahm established the first brewery in Davenport, Iowa.[83]

Beer gardens, associated with saloons or breweries, were an important part of the social scene in St. Louis by the 1850s.[84]

Figure AI.8. Bergner's Brewery in Philadelphia in 1857. (*100 Years of Brewing*, 1903, 199.)

1852 Frederick Seitz proposed boiling corn to a jelly before using it as a brewing adjunct.[85]

Henry Saxer started a brewery in the Portland, Oregon area; it was later owned by Henry Weinhard.[86]

1853 John Levesque of London published "Levesque's Improved Plan for a Brewhouse" which featured the recommended layout of the interior of a brew house. Gravity flow concepts were apparent.[87]

Yreka, California, saw the opening of its first brewery.[88]

1855 William Menger started a lager brewery in San Antonio, Texas, the city's first brewery.[89]

Lager beer riots occurred in Chicago over enforcement of Sunday closing laws.[90]

Emile Meier started a brewery in eastern Washington Territory at Walla Walla.[91]

"Know Nothings" rioted against Germans in Louisville in August and burned William Armbruster's brewery. Between twenty-four and two hundred persons were killed.[92]

1856 Henry Muench founded a brewery north of the Columbia River in Vancouver, Washington.

Fifteen breweries were in operation in San Francisco.[93]

1857 Eberhard Anheuser took control of the brewery that would someday be Anheuser-Busch, the largest brewing company in the world.

Cleveland was home to more than seventy-five lager beer saloons and gardens[94] (fig. A1.8).

Hop growing was introduced in California when Daniel Flint planted south of Sacramento and built the first hop kiln and first hop press on the West Coast.[95]

Henry W. Putnam of New York City patented fasteners to hold corks in beer bottles in 1857 and 1859.[96]

1859 Solomon, Tascher & Co. started Colorado's first brewery in Aurora.

New York grew 90 percent of America's hops.[97]

Messrs. Good and Endlich founded the first brewery in Denver. It was later known as the Phillip Zang Company.[98]

1860 Lager made up about one-fourth of American beer production.

The United States was home to 1,269 breweries.

An average brewery in New York state had eight employees. They were paid an average of $347 per year, with a range of $240 to $360 per year.[99]

The idea of using chip casks to clarify beer was introduced.[100]

Hop growing in Wisconsin began.

Eberhard Anheuser took over the bankrupt brewery of Hammer and Urban, the twenty-ninth largest of the forty breweries in the city of St. Louis.

In Idaho, the only out-of-state beers advertised during this decade were British Ale and Porter, both in bottles and on draft.[101] (Salt-glazed stoneware bottles from English breweries dating from 1805 to 1890 were later found throughout the West.)

In Nevada, Klein, Gerhauser and Wagner bought the Carson City Brewery that would remain in operation until 1948.[102]

1861 Wealthy brewer Matthew Vassar founded Vassar Female College.

John V. Hofstaetter hauled a six-barrel (186 gallon) brew kettle by wagon from The Dalles, Oregon, to Fort Colville, Washington, to start his brewery.[103]

Cincinnati brewers suffered reduced sales until the Union got control of the Mississippi River and they could increase exports downriver in 1863.[104]

1862 The newly formed Bureau of Internal Revenue began collecting an excise of $1 per barrel of beer brewed. Brewers were also required to pay $50 per year for their manufacturing license.[105]

The U. S. Brewers' Association—all members German-born—was organized in New York City as a result of the taxes imposed on brewers to finance the Civil War.

Federal tax records from this time on became a source for the name, location, ownership, and production amount for all legally operating breweries.

Ernst Weisgerber started the first brewery in the future territory of Idaho at Lewiston in a board-and-batten frame building.[106]

The modern era of brewing originated in the United States this year, according to the author of *One Hundred Years of Brewing.*[107]

1863 Thomas Smith, Christian Ritcher, and Henry Gilbert founded the first brewery in Montana at Virginia City[108] (fig. A1.9).

Figure AI.9. Gilbert Brewery, Virginia City, Montana, as of 1995. (Photo by author.)

Hauck and Windisch's Dayton Street Brewery in Cincinnati began construction of a new brewery in the Romanesque Revival style, very different from pre-Civil War breweries.[109] The style was also linked with the German *Rundbogenstil* architectural trend.

1864 Matheus Gottfried, a Chicago brewer, patented a device to use hot air instead of direct fire to pitch the inside of brewery barrels, vessels, and kegs.[110]

Chimarazo, the Confederate military hospital in Richmond, had a 400-keg brewery.[111]

Edward France, New York State, patented his Improved Hop-Dryer. The device removed dried hops from the kiln quickly by turning a lever attached to a shaft around which the cloth under the hops was attached.[112]

1865 Bressler proposed adding ground corn to the mash while the stirring apparatus was running.[113]

Samuel R. Percy and W. S. Wells patented a method for removing lupulin from hops and making it into a concentrate.[114]

1866 Hop growing in Washington Territory began (fig. A1.10).

Questions as to how to tax beer were settled with the rule that a stamp had to be attached to the spigot on every barrel removed from a brewery, effective September 1, 1866.[115]

1867 R. d' Heureuse, a Frenchman living in California, developed and patented a device to circulate air to germinating grain.[116]

P. H. Van der Weyde of Philadelphia patented a device to remove heat from water by use of a large variety of agents, such as gasoline, naphtha, condensed petroleum gas, and other substances.[117]

Figure AI.10. A hop yard at Puyallup, Washington. (Meeker, 1883, 52–53.)

A brewer from Germany traveling through America reported that first-class lager breweries in America were the equal of those in Munich and equal to second-class Vienna breweries. He described some small breweries as being near a spring, in a small house, with a diminutive brewing copper and mash tub. The hall of the house was the malting floor, the kiln was the size of a bake oven, and the cellar with fermenting tubs was four meters or less square.[118]

George Ehret opened his Hell Gate Brewery in New York and later reflected that, as with all breweries of that time, he had to do without "innumerable modern inventions and appliances" that had not yet been invented.[119]

Pabst Brewing quit burning wood and replaced it with coal, thus saving $3 per day.[120]

1868 Bernheimer Brewery of New York tried a patented method of mashing, in which adjunct grain is opened with steam before the malt is added.[121]

A. Schwarz established Der Amerikanische Bierbrauer in New York City.

The 1868 Revenue Act required wholesale liquor dealers, and brewers who sold from their warehouses, to pay a $100 special tax.[122]

Dr. J. E. Siebel established a laboratory in Chicago that became the Zymotechnic Institute.[123]

Boise, Idaho, brewer John Lemp added a stone cellar to his brewery.[124]

Jonathan Whitney of Port Hope in Columbia County, Wisconsin, patented his dumping hop box which ran on tracks from the kiln to the cooling room.[125]

1869 George Merz, a New Orleans brewer, installed the first mechanical refrigeration machine in the U. S. It used ether and was based on the Frenchman Charles Tellier's design and was not satisfactory.[126]

Completion of the first transcontinental railroad allowed eastern beer to be shipped into the West.

The first hop building in Oregon was built by William Wells at Santiam Bottom, Marion County, Oregon.[127] The state produced 9,745 pounds of hops that year.[128]

1870s Mr. De Coppet took out a number of patents on ice-making machinery that was manufactured by the Cincinnati (later Blymyer) Ice Machine Company.

1870 The United States was home to 1,972 breweries.

Pabst purchased the South Side Brewery in Milwaukee with its brew kettles, malt kilns, cellars, steam engines, dock on the Menomonee River, and railroad sidings.[129]

S. Liebmann's Sons Brewing of Brooklyn installed an ammonia ice-making plant from the design of Carré.[130]

Brewers had generally discontinued brewing lager in summer and top fermented beer in winter. This year, lager began to be brewed year round.

Franz Windhausen obtained a patent for his ice machine in the United States.[131]

"Ale breweries of the East still surpassed in magnitude of operation and equipment the lager beer breweries."[132]

1871 Horace Greeley and others published the 1,304-page *The Great Industries of the United States* in which he attacked beer drinkers.

The first works for hop extraction was erected at Long Island City, New York, by the New York Hop Extract Company.[133]

George B. Cornell of Chicago patented an iron screw bushing for the protection of bungholes of beer barrels.[134]

At least eight breweries were destroyed in the Great Chicago Fire.[135] Bob Skilnik's book listed nineteen breweries worth over $2,000,000 lost.[136]

John Flintoff established the *American Brewers Gazette* in New York.

1872 Revenue Act declared anyone who sold malt liquor in quantities over five gallons to be a wholesale dealer subject to a special tax of $50.[137]

The first known strike at an American brewery occurred in New York in May as part of a general strike.[138]

1873 English replaced German as the official language of the U. S. Brewers' Association.

Adolphus Busch, having perfected pasteurization, began bottling beer for large-scale shipment.[139] By 1872 or 1873 the brewery had eighty employees and two stories of automated bottling equipment turning out 40,000 bottles per day.

Midwestern brewers considered $10 per barrel a good price for beer.[140]

The total number of breweries in the United States reached 4,131—the largest total ever.[141]

Pabst Brewing bought 180,000 bushels (440 carloads) of barley from Italy.[142]

1874 Adolphus Busch introduced refrigerated railcars about this time, ahead of meatpacker Gustavus Swift, who has often been called the originator of this technology.[143]

1875 Franz Windhausen designed the ice-making machine installed in the Christian Moerlein Brewery in Cincinnati.[144]

The malting floor developed in Germany by Mr. Weinig was first used in the United States.[145]

The first cork-free breakthrough for closing beer bottles was the lightning stopper patented by Charles De Quillfeldt of New York City in 1875.[146]

1876 The Centennial Exposition in Philadelphia featured machinery and materials for brewing. Hops from Napa County, California won an award for the best hops.[147]

Philadelphia hotels sold large amounts of lager beer during the Exposition, hurting the sale of "fancy drinks" and imported beers.[148]

Louis Pasteur published his studies of beer, showing that application of heat kills bacteria and helps preserve food products. Adolphus Busch led the way in applying this knowledge to American beer.

New York Hop Extract Company relocated to Waterville, Oneida County, New York.[149]

Vaas and Littman's improved version of Carré's ice making machine was brought to Cincinnati and used in a brewery there for four or five years.[150]

A. T. Ballentine installed his ice machine in a small New York brewery and it was an utter failure.[151]

The Western Brewer established in Chicago as English-language brewing periodical.[152]

Fred Baumann, architect, laid down the general plan of a modern brewery in the November issue of *The Western Brewer.*[153]

1877 George Ehret of New York led American brewers with 139,000 barrels brewed.[154]

Bergner & Engel Brewery in Philadelphia first installed a refrigeration machine designed by Daniel L. Holden.[155]

David Boyle installed his refrigeration machine, using a compression ammonia plant, in the Bemis & McAvoy Brewery in Chicago.[156]

Silas Merchant ammonia compressor ice machine installed in the brewery of A. Ziegel, Buffalo, New York.[157]

Rudolph Schmidt, a St. Louis malt manufacturer, installed a refrigerating machine designed by Raoul Pierre Pictet of Switzerland.[158]

Lemp of St. Louis had a bottling house including cooper shop and stave yard that took up an entire city block.[159]

The eight largest brewers in the United States began a period of enormous growth up to 1895. For instance, Anheuser-Busch grew 1,106 percent and the industry average was 242 percent.[160]

1878 Charles Kaestner of Chicago took out his first patent on the many improvements he had made on mashing equipment.

Ferd Heim Brewery in Kansas City installed a Boyle ice machine that was successful.[161]

Lemp Brewery of St. Louis installed a refrigeration machine of the ammonia compression type developed by Theodore Krausch.[162]

Anheuser-Busch began using the new J. M. Ayer Rubber Refrigerator Cars for railroad transportation of its beer.[163]

Danish chemist E. C. Hansen developed the theory of "pure" yeast.[164]

1879 The Continental Brewery in Philadelphia was the first in the world to be lighted by electricity.

Thomas L. Rankin constructed his first refrigeration machine that was initially installed in the Jacob Ruppert Brewery of New York.[165]

Hermann Brewery of New York installed the first De La Vergne refrigerating machine.[166]

A stone brewery was built in Challis, Idaho, that stood into the twenty-first century and was entered on the National Register of Historical Places.[167]

After a tour of the country's larger breweries, Pabst Brewing installed a new style of steep tank for malting.[168]

Brewer John Hepburn reported a "large beer boiler that was taken into Bonanza [Idaho] over the Cape Horn trail did not receive a dent. It was the most difficult piece of freight to carry that has been taken into Yankee Fork this season." [169]

1880 The United States had 2,272 breweries in operation, a large decrease since 1873.[170]

A pneumatic malting plant, as designed by Nicholas Galland in France, was installed in the malt plant of Neidlinger, Schmidt & Company in New York City.[171]

Pabst installed its new Boyle ice machine in the Empire Brewery.[172]

Hop growing began in Oregon. New York produced over 80 percent of the American hop crop.[173]

Dr. A. Schwarz established a scientific station to study brewing in New York, and the first brewer academy in the United States.[174]

The first commercial air-cooling unit in America was installed in the brewery of Robert Portner in Alexandria, Virginia.

David W. Davis patented an icehouse that used salt to make the ice colder. E. W. Voigt of Detroit installed the first one at his brewery[175] (fig. A1.11).

Wacker & Birk Brewing and Malting of Chicago was first to install a type of refrigeration plant designed by C. P. G. Linde of Munich and manufactured in America by the Fred W. Wolf company of Chicago.[176]

About this time, a marked change in brewery construction was inaugurated due to mechanical refrigeration and the discarding of many cumbersome devices to store and use natural ice. *The Western Brewer* recorded these changes:[177] ". . . the whole picture of the brewery, the building fronts and the entire exterior architecture aroused laymen and professional men to a sense of the magnitude of the business enterprise, of the success hitherto achieved, and possibilities for the future, of so flourishing an industry."[178]

1881 Nicholas Galland patented in the United States his invention for germinating grain.[179]

W. Toepfer & Sons built the first successful dumping malting kiln floor in Milwaukee.[180]

Brewery worker strikes in New York and Cincinnati sought twelve-hour workdays and two hours of work on Sunday. The strikes failed.[181]

1882 Louis C. Huck, a Chicago maltster, took out a patent on a system to air-dry malt and ventilate malting houses.[182]

E. W. Voight brewery of Detroit installed a patented "Non Plus Ultra Ice-House" that used natural ice with salt to obtain a low temperature.[183]

Philip Best Brewing Company of Milwaukee installed the first large malting kiln of the Toepfer dumping design.

A. T. Ballentine installed an ammonia-compression ice machine in the brewery of Gerhard Lang of Buffalo, New York.[184]

Cream City Brewery of Milwaukee installed a refrigeration machine designed by Peter Weisel and built by Wesiel and Vilter Company.[185]

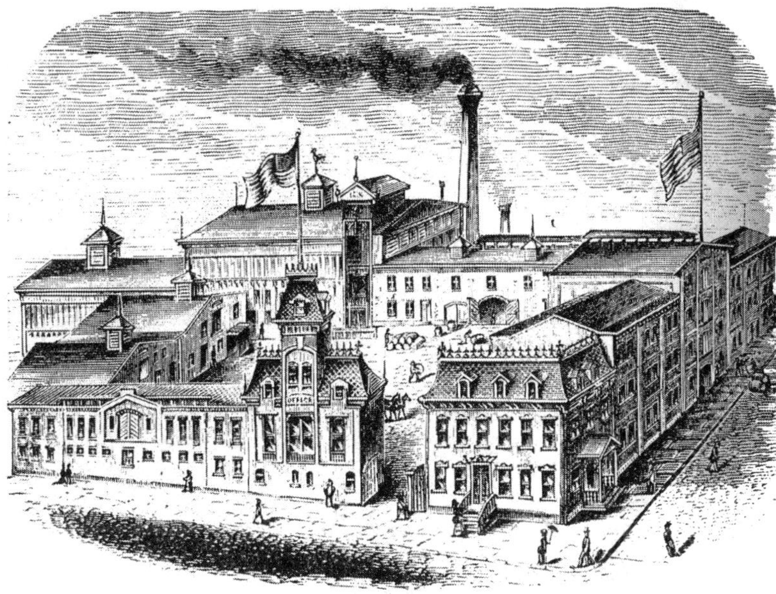

Figure AI.11. Kruger Brewery, New Jersey, 1880. (Salem [1880] 1972, 162–163.)

Anheuser-Busch had three 50-ton Boyle ice machines in operation, the largest capacity mechanical refrigeration plant in the world.[186]

1883 About this time, brewers used separate vessels called rice tubs or cookers to prepare adjuncts before adding them to the mash.[187]

1884 Goulding bottle-washing machines were introduced.[188] Vilter Manufacturing developed bottle-filling machines that superseded the rubber hose.

Brewer's Union No. 1 was formed in New York as part of the Knights of Labor Local Assembly 1672, thus beginning trade unionism in the modern brewery industry.[189] Ale and porter workers had their own union that was English, not German.

1885 Only one beer bottle factory west of the Mississippi was reported. Western brewers reused empty bottles belonging to eastern brewers.[190]

1886 Use of pure culture yeast began to be generally adopted by American brewers.[191]

Abraham Cohen established the first brewery in Alaska at Juneau.[192]

Henry Werner's Phoenix Brewery in Baltimore first installed a David Smith "Eclipse" refrigerating machine (a 20-ton size) made by the Frick Company.[193]

Drs. Wahl and Henius established a scientific station to study brewing in Chicago.[194]

The Brewers', Beer Drivers', and Maltsters' Unions successfully negotiated a contract with the Brewers Association in the New York metropolitan area for a ten-hour day, six days per week, with a weekly wage of $15 to $18.[195]

The *Brauer-Zeitung* [Brewer's Newspaper] was founded in the German language for brewery union members.[196]

John Zynda founded the first Polish brewery in Detroit.[197]

Brewery workers in Chicago negotiated a 64-hour work week with free beer five times per day but only three glasses each time.[198]

1887 U. S. Brewmasters' Association was organized. Its publication was written in German.

Michigan Brewers' and Maltsters' Association held its first convention.[199]

Goetz & Flodin dumping kiln floor was first introduced after being patented the previous year.

Knoxville Brewery of Tennessee installed a refrigeration machine manufactured by the Henry Vogt Machine Company.[200]

Pabst introduced Stockheim filters to America to replace chip casks for "fining" or clarifying beer.[201]

Pabst introduced pure yeast culture in its brewery.[202]

1888 William Heiser of Buffalo, New York, patented a manhole door for beer storage vats.[203]

The United States grew 44 million bushels of barley.[204]

Bullock Electrical Manufacturing Company of Cincinnati installed electrical equipment for the first time in the George Wiedemann Brewery of Newport, Kentucky.[205]

Pabst Brewing ceased to use any natural ice.[206]

In New York, brewery owners locked out union members. Four thousand laborers were affected.[207]

1889 J. F. Theurer, Pabst superintendent of brewing, developed a system to collect carbon dioxide, clear it of impurities, and inject it into the finished beer.[208]

New nozzles and lower temperatures at the Pabst facilities enabled the company to reduce the shrinkage in the volume of beer between storage tank and shipping container from 22.18 percent in 1886 to 10 percent in 1889.[209]

1890 The number of breweries in the United States declined to 1,928.

Adolph Coors of Golden, Colorado, became a millionaire.

Pabst Brewing began to dry spent grain before sending it out for animal feed.[210]

Congress altered the Internal Revenue Act so that beer could be piped from storage cellar to bottling house instead of barreled and hauled over a public road.[211]

1891 Windhausen patented an ice machine that used carbonic acid gas as the cold-producing medium.[212]

The eight-story malt house at Pabst in Milwaukee did traditional floor malting on the top six floors but had a Saladin pneumatic plant of sixteen compartments on the two lowest floors.[213]

Boise, Idaho, brewer John Lemp bought an 8,105-pound "refrigerator" from the Brunswick-Balke-Callender Company of Chicago.[214]

1892 F. B. Giesler invented a dumping malt kiln floor that was manufactured by Galland-Henning Company of Milwaukee.

The cork-lined, 21-fluted, metal Crown cap was invented by William Painter of Crown Cork and Seal Co. in Baltimore.[215]

Robert Smith invented the wood pulp coaster in Dresden, Germany.[216]

Adolphus Busch earned an annual salary of $75,000; Anheuser-Busch earned a $750,000 profit.[217]

The Rothschilds of England sought to combine Lemp and Anheuser-Busch of St. Louis with Schlitz and Pabst of Milwaukee into a $10 million cartel. The plan failed.[218]

Pennsylvania Iron Works made and installed a Boyle compression ice-making machine in the brewery of J. W. Russert in La Porte, Indiana.[219]

1893 J.A. Tilden patented a drum malting system. He got other patents on the device in 1897 and 1898.

Bottled beer became 10 percent of the sales of Pabst Brewing.[220]

At the World's Fair in Chicago, Bartholomay Brewing Company of Buffalo installed a four-foot-square working model of the interior of its brewery complete with ice machine and mash tub.[221] Pabst put up a gold-plated 13-foot-square model of his brewery.

1894 W. H. Prinz of Chicago patented the automatic and pneumatic malting kiln.[222]

Ninety-five percent of the 852,000 barrels of beer consumed in Cincinnati was locally brewed.[223]

1895 Albert Schwill & Company of Chicago first used the pneumatic drum malting system developed by Stier.[224]

C. M. Warner installed the first Dornfield complete ventilating and compartment malting system at Syracuse, New York.[225]

From 1895 to 1915, all six national shippers grew more slowly than the brewing industry as a whole.[226]

1896 The United Brewery Workers Union held its ninth convention in Cincinnati. They had 104 local unions and 8,072 members.[227]

1897 $250 million in capital was invested in breweries; beer production annually was valued at $250 million.[228]

1898 To finance the Spanish-American War, the tax on beer was raised.[229]

Oregon passed New York to become the largest hop producing state.[230]

The United States grew 90 million bushels of barley, an enormous increase.[231]

1899 Pabst pioneered advertising with matchbooks by ordering 10 million.[232]

Oregon Hopgrowers' Association was incorporated to control output and prices.[233]

1900	The number of breweries in the United States continues to decline to 1,816 breweries in operation. Some sources record only 1,509. Approximately 19,900 brewery workers were members of unions, with 202 local unions and 83 branches. The industry paid $25,826,211 in wages to 39,532 workers, including 504 women and 643 children.[234] The United States was home to 146 independent maltsters.[235] Brewers paid over $73 million in federal taxes.[236] Brewer John Brodbeck bought and installed the first ice-making equipment in the city of Boise, Idaho.[237]
1901	First motor truck, "Our Mary Ann," was used by an American brewery at the Central Brewing Co. of New York City.[238]
1902	Beer tax returned to $1 per barrel. Brewery workers of Cincinnati were involved in a protracted strike.[239]
1903	Automatic bottling machine came into use.[240]
1905	The number of licensed breweries in the United States was recorded as 1,847, the highest total for the twentieth century. Breweries that brewed more than 170,000 barrels produced 28.2 percent of the nation's beer. Breweries that produced 15,000 to 170,000 barrels per year produced 62.3 percent of the nation's beer.[241] In terms of capital invested, brewing was the third largest industry in the United States.[242]
1906	The San Francisco earthquake destroyed six breweries and damaged twenty of the twenty-four other breweries in the city.
1907	The new stock house at Hamm's of St. Paul was the ninth building and the first brewery structure in St. Paul to use a new system of reinforced concrete.[243] The building was sleek, modern, and free of the Romanesque details popular a few years earlier.
1908	The Idaho Falls Brewing Company built a new malt house that had a 50,000-bushel grain elevator beside it. The malting company was a separate corporation but with the same stockholders as the brewery.[244]
1909	A Montana brewer contacted American Can Company to see if cans were available for packaging beer.[245] E. Clemens Horst, a California hop grower, invented a stationary hop-picking machine.[246]
1910	The United States records 1,568 breweries. New York produced more beer than any other state. Rockwell Clough of Alton, New Hampshire, patented a combination bottle opener and corkscrew named "the Decapitator." In August, one of the largest forest fires in North American history swept into Wallace, Idaho, burning the Sunset Brewery and destroying two thousand barrels of beer.[247]
1911	Grangeville, Idaho, brewer and grain grower William Von Berge won the prize of the National Brewers and Hop Growers Association for the best barley in the United States.[248] The United States produced more beer than any other nation for the first time.[249] The Aluminum Company of America advertised in brewery trade magazines for the first time that aluminum remained neutral to all brewing materials and was an ideal metal for brewers.[250]
1912	Stroh Brewery of Detroit adopted "fire brewing" with direct fire under the kettles instead of heating them with steam.[251]
1913	Sake brewery started in Hilo, Hawaii. Jacob Rupert Brewery in New York City had 126 motor vehicles to lead the nation; Anheuser-Busch of St. Louis had ninety-two, and George Ehret's Hellgate Brewery of New York had eighty-three.[252]
1917	A book entitled *New Fields for Brewers* recommended manufacture of near beer, fruit juice, yeast, malt extract, vinegar, breakfast foods, dairy products, industrial alcohol, and mechanical appliances to replace brewing.[253] Tax on beer was raised to $3.00 per barrel in October.[254] Beer was limited to 2.75 percent alcohol by weight on December 11.[255]
1918	On September 16, President Wilson forbade the use of any grain for brewing malt liquor.[256]
1919	Nebraska became the thirty-sixth state to ratify the 18th Amendment on January 16, meaning it would be the law of the land exactly one year later. The Volstead Act, designed to enforce the 18th Amendment, passed over President Wilson's veto. This legislation made anything with more than one-half of 1 percent alcohol an illegal alcoholic beverage.[257]

American hop growers got high prices—one dollar per pound—due to the untended fields of Europe during the World War.[258]

Beer tax was raised to $6.00 per barrel.[259]

1920 The 18th Amendment went into effect.

1921 The five hundred breweries in the United States had to produce something other than beer or shut down.

Near beer was produced for a total of 300 million gallons.[260]

Anheuser-Busch put Budweiser Barley Malt Syrup on the market.[261]

The Willis-Campbell Bill outlawed medical prescriptions for beer and extended Prohibition to the Virgin Islands and Hawaii.[262]

1922 Lemp Brewery in St. Louis was sold to the International Shoe Company.[263]

Five hundred breweries were producing near beer.[264]

1924 Joseph Stenson ran the Manhattan, Stege, Pfeiffer, Standard, Gambrinus, and Hoffman breweries for gangster Johnny Torrio in Chicago. They sold beer for $50 to $55 per barrel and earned $12 million.[265]

1925 This was the last year in which significant amounts of winter ice were cut and stored. Electric refrigerators made the business obsolete.

Anheuser-Busch marketed hopped malt syrup for making "cookies."[266]

Montana became the first state to repeal its Prohibition enforcement laws.[267]

1927 Budweiser Yeast entered the U. S. market.[268]

Hop-pickers in the Northwest were from many states, "a sprinkling from Canada and Mexico, added to Japanese, Chinese, Russians, Filipinos, Hawaiians and sometimes Alaskans."[269]

1930 Pabst Brewing began to acquire new storage tanks, kegs, and other equipment in anticipation of the end of Prohibition.[270]

1931 Near beer production reached 97,243,528 gallons. This was one-third of the total of 1921.[271]

1932 American Can Company developed a new type of opener for cans filled with liquid; it was named the "Quick and Easy."[272]

1933 Legal beer returned to the U.S. The Cullen-Harrison Bill (Beer-Wine Revenue Act) made 3.2 percent beer legal. By the end of the year, the 21st Amendment repealed the 18th Amendment.[273]

1934 Food City Beer, brewed by the Food City Brewing Co. of Battle Creek, Michigan, became the first to be brewed in glass-lined brew kettles.

American Can Company trademarked "Keglined" for its interior coating for beer cans on September 24, 1934.[274]

About this year, Anheuser-Busch rediscovered the abandoned and forgotten natural caverns beneath the brewery in St. Louis.[275]

Bottled beer made up 25 percent of beer sales.[276]

On April 1, one year after the repeal of Prohibition, 669 breweries were licensed to operate.[277]

Aztec Brewing Co. of San Diego used sky-writing to advertise.[278]

American Brewers used 1,433 million pounds of barley, 253 million pounds of corn, 103 million pounds of rice, and 26 million pounds of hops.[279]

1935 American Can Company and Gottfried Krueger Brewing Company of New Jersey marketed the world's first beer can.[280] By year's end, twenty-three brewers canned their beer. Cans usually had instructions with illustrations for opening the can. After June, all cans read "Internal Revenue Tax Paid." Some states required cans to have the permit number of the brewing license of the brewery.

Schlitz introduced the cone top beer can.[281] The earliest ones had flat bottoms and short cones, called low profile.[282] Existing bottle lines in breweries could fill these with only minor modifications.

Falstaff Brewing of St. Louis leased the Krug Brewing Company of Omaha and touched off a wave of acquisitions by large brewers.[283]

John F. Betz & Son Brewery of Philadelphia bought the old American Theatre and transformed it into a bottling house.[284]

Owens-Illinois Glass Co., the leading manufacturer of beer bottles, introduced the no deposit/no return "Stubby" bottle. It was 3 inches shorter than traditional beer bottles to compete with cans.[285]

1936 January 24 marked the first anniversary of the commercial use of the beer can. American Can Company was supplying twenty breweries, Continental supplied twelve, and National sold to five.[286]

Downy mildew destroyed much of the hop crop, causing prices to rise to 29 cents per pound.[287]

The United Brewers Industrial Foundation began, with the goal to change public attitudes toward beer.[288]

1937 The 720 breweries in operation in America produced slightly less beer than the 1,300 breweries of 1914.[289]

Bottled beer made up 44 percent of all beer sales.[290]

Cone top cans known as the J-spout were introduced by Crown Cork & Seal Company.[291] Crown became the second cone top manufacturer.[292]

Crown Cork & Seal Company produced 8-ounce cans for Peter Fox Brewing Company of Chicago.[293]

Edward Thys invented a portable hop-picking machine that was towed through the fields with a tractor.[294]

1938 Crimped and glued seams, often called toothed seams, first appeared on 32-ounce cans.

1939 Oregon produced half the nation's hops. Washington produced one-fourth of the crop, as did California.[295]

1940 The Federal Alcohol Administration ruled that vitamins could not be mentioned on beer packages.[296]

Blight in this decade ended New York hop growing.[297]

J-spout cans were phased out and replaced by the "crowntainer" type that was smooth and coated with aluminum.[298]

There were 598 active breweries in the U.S.[299]

American brewers used 1,958 million pounds of barley, 441 million pounds of corn, 189 million pounds of rice, and 32 million pounds of hops.[300]

1941 Canned and bottled beer made up 51.7 percent of sales, passing kegs and barrels at 48.3 percent for the first time.[301]

Only 186 of the 507 breweries in the United States were canning beer.[302]

The pop top bottle caps, with no opener needed, were briefly popular.[303]

Over one billion beer cans were made for the first time.[304]

1942 Canning of beer was prohibited to save metal for the war effort.

1943 Brewers were required to allocate 15 percent of their beer for the military.[305]

Military beer cans were silver or olive drab in color, and did not read "Internal Revenue Tax Paid" but "Withdrawn Free of Tax for Exportation."[306]

Barley malt was rationed due to the war.[307]

1944 The Army Quartermaster Department designed a 12-ounce beer bottle that was 50 percent lighter than the pre-war bottle; 426 million were shipped overseas.[308]

The United States Brewers Association merged with the United Brewers Industrial Foundation to become the new United States Brewers Foundation.[309]

1945 Fifty million cases of beer were sent overseas to U.S. troops.[310]

German prisoners of war from the Corvallis area harvested Oregon hops.[311]

Pabst purchased the Hoffman Brewery in Newark in December.[312]

American brewers used 2,199 million pounds of barley, 940 million pounds of corn, 215 million pounds of rice, and 37 million pounds of hops.[313]

1947 The brewing industry voluntarily set up a grain conservation program to help feed starving Europe.[314]

1948 Pabst bought control of the Los Angeles Brewing Co.[315]

Goebel Brewing Co. of Detroit distributed its beer nationwide in 7-oz. bottles.[316]

1949 The U. S. Third Circuit Court of Appeals gave DuBois Brewing Company of Pennsylvania the right to continue to use the name Budweiser on its beer.[317] Anheuser-Busch had sued companies that used this name many times.

1950 Jos. Schlitz Brewing Company donated 600,000 cans of beer to U.S. troops in Korea.[318]

Miller High Life achieved distribution in all forty-eight states.[319]

Eighty-five percent of the hops grown in America were picked by machine.[320]

The statement "Internal Revenue Tax Paid" was removed from can and bottle labels.[321]

Returnable bottles were 78.5 percent of packaged beer sales, cans were 18.9 percent, and nonreturnable bottles were 2.6 percent.[322]

American brewers used 2,707 million pounds of barley, 852 million pounds of corn, 321 million pounds of rice, and 38 million pounds of hops.[323]

The five largest brewers in the U.S.A. were: Schlitz with 5.1 million barrels and 6 percent of the beer market; Anheuser-Busch with 4.9 million barrels and 6 percent; Ballentine with 4.4 million barrels and 5 percent; Pabst with 3.4 million barrels and

4 percent; and Schaeffer, whose 2.8 million barrels made up 3 percent of the market.[324]

1951 Anheuser-Busch built a new brewery in Newark, N.J., which began a trend for large breweries to expand by building new regional breweries.[325]

1952 American brewers used 2,654 million pounds of barley, 920 million pounds of corn, 337 million pounds of rice, and 35 million pounds of hops. In the eighteen years since the repeal of Prohibition, the ratio of corn in beer rose from 14% to 23%, rice rose from 5½% to 8½%, and barley dropped from 79% to 63%.[326]

Cans constituted 24 percent of packaged beer sales.[327]

Fewer than four hundred hop growers operated in Oregon; this was down from more than one thousand growers in the late 1940s.[328]

1953 Theo. Hamm Co. of Minneapolis purchased the San Francisco plant of the Rainier Brewing Company and rebuilt the plant from the bottom up.[329] The Rainier name was sold to Sicks' Seattle Brewing and Malting Company for an estimated $500,000.[330]

1954 The U. S. Senate Small Business Committee released a study predicting that a decade later, ten brewing companies would control 70 percent of total beer sales. More than two hundred small breweries were for sale at the time.[331]

Edward J. Griesedieck, president of the Griesedieck Bros. Brewing Co. of St. Louis, donated $1,000 to the Missouri Historical Society for a brewing industry historical collection.[332]

Schlitz marketed the 16-oz. punch-top can.[333]

1958 Primo of Hawaii marketed an 11-oz., paper labeled, all-aluminum can.[334]

Brewery worker strike in Detroit shattered Stroh's growth.[335]

1959 Coors introduced all-aluminum beer cans in a 7-oz. size.[336] These were retailed in eight-packs.

Beer sales set a record, breaking the totals for 1947. Packaged beer exceeded 80 percent of sales for the first time.[337] Canned beer outsold bottled beer for the first time.[338]

1960 Aluminum tops for beer cans were introduced.[339]

The United States had 229 active breweries. Wisconsin had thirty-three and Pennsylvania had twenty-six.[340]

The last cone top cans were filled by the Rice Lake Brewing Company.[341]

The five largest brewers in the U.S.A. were: Anheuser-Busch with 8.5 million barrels and 10% of market share; Schlitz with 5.7 million barrels and 6%; Falstaff with 4.9 million barrels and 6%; Carling with 4.8 million barrels and 5%; and Pabst with 4.7 million barrels and 5%.[342]

1962 Tab top, pull tab, or snap-top can was introduced by Pittsburgh Brewing Company.[343]

Short skirt, plastic-backed crowns for bottles appeared; the older style crowns disappeared.[344]

1963 Schlitz introduced tab top cans on a national basis.

The first 12-oz. all-aluminum cans were issued.[345]

Plastic six-pack holder (yoke) was introduced.[346]

Washington State produced more than half the nation's hop crop.[347]

1964 In St. Louis, Falstaff opened the first museum in the United States dedicated to the brewing industry.[348]

Sixty percent of all beer cans used tab tops.[349]

Gallon cans, tab tops with smile beads, and Continental Can's "U-Tab" design were introduced.[350]

1965 Ring-pull beer can was introduced.[351]

Hamm's beer bear won the number one ranking as the best-liked advertisement in the U.S. despite being viewed in only thirty-one states.[352]

1966 Welded-seam cans were introduced by Continental Can. American Can made cans with tin-free steel (TFS) with cemented "Mira-Seam" seams.[353] The "neck-in chime" can with a lid smaller than the body was introduced.[354]

1967 Anheuser-Busch's original St. Louis plant became the first brewery to be a Registered National Historic Landmark.[355]

Tin-free steel cans were introduced into wide distribution.[356]

Schlitz patented accelerated batch fermentation (ABF). By injecting air into a brew, the yeast grew faster and fermentation was quickened by ten days.[357]

1969 Canned beer outsold bottled beer for the first time.[358]

Fritz Maytag bought Anchor Steam Beer of San Francisco and began a trend toward high quality non-mainstream beer.[359]

1970s Production of 11-oz., 15-oz., and gallon cans ended.[360]

UPC computer codes were introduced on cans, bottles, and six-pack holders.[361]

1970	Two-piece can construction developed.[362]
	The United States had 142 active breweries.[363]
	Cans commemorating local and national events became widely distributed.[364]
	The top five brewers in the U.S.A. were: Anheuser-Busch with 22 million barrels and 18% of market share; Schlitz with 15 million barrels and 12%; Pabst with 10 million barrels and 8%; Coors with 7 million barrels and 6%; and Schaeffer with 6 million barrels and 5% of market share.[365]
	Beer Can Collectors of America was formed. Eastern Coast Breweriana Association was established.
1971	Twist-off crowns were developed and put into use the following year.
1972	Miller Brewing bought the "Lite" brand and began the trend to light beer.[366] Parent company Phillip Morris had funds for advertising.
	Can lines in breweries process 1,200 per minute and bottle lines process 900 per minute.[367]
	Oregon banned ring-pull cans. Coors introduced push-button opening cans. Cans with specialized shapes were introduced.[368]
	National Association of Breweriana Advertising was formed.
1975	American Can Company began producing push-button cans.[369]
	Olympia of Tumwater, Washington, bought Hamm's Brewing.[370]
1976	The U.S. produced more than twice as much beer—163.8 million barrels versus 81.5 million barrels—as the runner up nation, West Germany.[371]
	Olympia Brewing merged with Lone Star.[372]
	Schlitz dumped 10 million cans of Memphis and Tampa output because the beer was hazy.[373]
1977	Jack McAuliffe founded New Albion brewery in Sonoma, California, and began America's Micro Brewery Revolution.[374] He had to design and build most of his equipment.
	Billy Beer named for President Carter's brother was introduced. A myth several years later credited the cans with having great financial value.
	Coors phases out push-button cans.[375]
1979	Home brewing of up to 200 gallons per household became legal.[376]
1980s	UPC codes became standard on all cans.
	Multiple neck-in chimes used in the early years of the decade on beer cans were changed to single, longer neck-in chimes later in the decade.[377]
1980	U. S. brewers brewed almost 150 million barrels of beer and paid the government 1.5 billion in taxes.[378]
	The top five brewers in the U. S. were: Anheuser-Busch with 50 million barrels and 28% of market share; Miller with 37 million barrels and 21% of market; Pabst with 15 million barrels and 9%; Schlitz with 15 million barrels and 9%; and Coors with 14 million barrels and 8% of market share.[379]
1981	The Lone Star Brewery in San Antonio, Texas, became the San Antonio Museum of Art.[380]
	Schlitz in Milwaukee closed.
	A large plant of the Western Malting Company was built in Pocatello, Idaho, at a cost of $32.5 million.
	Heileman and Pabst fought to buy Schlitz. Stroh's bought Schlitz the next year.[381]
1982	Bert Grant of Yakima, Washington, brewed and sold his beer along with food at his establishment. The American brewpub was born.[382]
	American Breweriana Association was formed.
1983	Fifty-one brewing concerns operated eighty breweries. This was the smallest number of breweries in the U.S. in the twentieth century.[383]
	Production of ring-pull cans ceased.[384]
1984	Forty-four brewing concerns operated a total of eighty-three breweries in the U.S.[385]
	At least eight new microbrewery startups indicated this trend was growing.
	Production of straight-sided steel cans ceased.[386]
1985	Paul Kalmanovitz, owner of many breweries, made a hostile takeover of Pabst for $63 million.[387]
	Stroh's closed its original brewery in Detroit and eventually tore it down.[388]
1987	In Challis, Idaho, the Klug brewery was excavated.
	Heileman of LaCrosse, Wisconsin, merged with the Australian Gold Bond Corporation.[389]
	The Oldenburg Brewery in Ft. Mitchell, Kentucky, near Cincinnati, opened with the Herb and Helen Haydock breweriana collection on display.

1988 Archaeological work at Harper's Ferry, Virginia, led to investigation of the brewery there.

"Dry" beers were introduced by major American brewers after Asahi Brewing of Japan introduced them there. Fermentation is allowed to progress further, leaving less residual sugar and less aftertaste.

1989 Government alcohol warning labels were introduced.[390]

1990 The RECON Company of San Diego made an archaeological investigation of the old Aztec Brewery in San Diego. The site in the 2200 and 2300 blocks of Main Street was the home of the Savage Tire Factory beginning in 1911. In the early 1930s, the complex was redesigned as a brewery.

G. Heileman Brewing of LaCrosse, Wisconsin, filed for bankruptcy.

The top five brewing companies in the U.S. were: Anheuser-Busch with 80 million barrels and 42% of market share; Miller with 61 million barrels and 22%; Stroh's with 21 million barrels and 10%; Coors with 20 million barrels and 9%; and Heileman with 19 million barrels and 8% of market share.[391]

1993 The Bureau of Alcohol, Tobacco and Firearms decided that "All therapeutic claims regardless of truthfulness, are inherently misleading and particularly deceptive," and would not allow Yakima Brewing and Malting Co. to list all nutritional information on its six-packs of Scottish Ale.

Archaeological investigation was made on the Shuter's Hill Brewery in the West End of Alexandria, Virginia.

1994 The printing of alcohol content on beer containers was legalized.[392]

Anheuser-Busch bought 25 percent of Red Hook of Seattle.[393]

"Ice" beers were introduced by major brewers. Slightly higher in alcohol, the beers were stored at about 28 degrees, allowing ice crystals to form which were later removed.

1995 Approximately five hundred breweries were operating in the U.S., with three or four opening every week.[394]

1996 Microbrew (craft beer) growth slowed from a 51 percent increase in 1995 to a 25 percent increase in 1996.[395]

Pabst closed its Milwaukee brewery.[396]

After Heileman Brewing's second bankruptcy, Stroh's bought its former competitor.

1998 Craft beer sales leveled out at 3 percent of the total American beer market.[397]

The largest selling brewpub in the U.S., Brew Brothers in the Eldorado Hotel and Casino in Reno, Nevada, sold 6,108 barrels of beer in 1998.[398]

Anheuser-Busch ownership in Mexican brewer Grupo Modelo S.A. de C.V. rose from 37 percent to 50.2 percent.[399]

Pabst Brewing Co. purchased Stroh's Brewing and closed most of the breweries.[400] Miller Brewing was involved in the three-way deal.

October floods damaged breweries in the Kansas City area of Missouri and Kansas.[401]

O'Brien Brewing and Malting in Klondike City, Yukon, Canada, was investigated by archaeologists from Simon Frazier University.

1999 The beer industry contributed $187.1 billion to the U.S. economy.[402]

2000 Coors accidentally spilled 2,500 barrels of beer into Clear Creek near the brewery and killed 50,000 fish.[403]

Pub Brewing Company manufactured, installed, and trained the users of brewing equipment for 7-barrel to 50-barrel breweries.[404]

Miller Brewing Co. became the first to have widespread distribution of recyclable plastic beer bottles.[405]

The largest breweries in the United States were: Anheuser-Busch with 98,300,000 barrels and 53.42% of market share; Miller with 41,229,000 barrels and 22.60% of market share; Coors with 22,994,000 barrels and 12.50% of market; Pabst with 10,500,000 barrels and 5.70%; and Genesee with 1,350,000 barrels or 0.07% of the market share.[406]

The home of brewer Henry S. Gilbert was investigated archaeologically in Virginia City, Montana (fig. A1.12).

2001 Guinness imported beer cans that had a self-activating "widget" to produce instant draft beer when the can was opened.[407]

Alaska-based manufacturer Incan developed a heat sealed, zip-close foil pouch to replace beer bottles and cans.[408]

Bud Light became the top-selling beer in the U.S., but Budweiser was still the fastest selling in the world.[409]

Figure AI.12. Structure near the home of Henry Gilbert in Virginia City, Montana. (Photo by author.)

Anheuser-Busch donated $20,000 to the Thomas Jefferson Foundation to restore the cellar at Monticello. The project included information about brewing at Jefferson's home.[410]

Anheuser-Busch removed the "shroud," the decorative paper covering over the top of the bottle, from the Michelob brand after forty years of tradition.[411]

Kreische Brewery in Fayette County, Texas, was assessed archaeologically.

2002 Anheuser-Busch stock rose 7 percent while the Dow Jones Industrial Average dropped 17 percent.[412]

Philip Morris Corp. sold Miller Brewing to South African Breweries for $5.6 billion.[413] The combined company was renamed SABMiller PLC.

TrenStar of Greenwood Village, Colorado, began leasing stainless steel kegs to breweries. The company picks them up empty, cleans them, stores them until the brewer needs them, and then tracks their location electronically to prevent theft.[414]

Minnesota Brewing Co., one of the last old-time regional breweries, closed down. Grain Belt was its most famous brand.[415]

Coors bought U.K.-based Carling for $1.7 billion. Included were British brands Caffrey's, Stones, and Worthington, totaling 19 percent of the U.K. market.[416]

Pabst closed its last two breweries but remained the fourth largest U.S. brewer, with beer brewed under contract by Miller.[417]

2003 The total U.S. beer market was 204,333,000 barrels of beer.[418]

Miller Brewing workers in North Carolina complained that the company had twenty-nine robots to move cases and barrels of beer, and the number of workers had been cut in half over the last twenty-five years.[419]

Miller closed its brewery in Tumwater, Washington, the former Olympia brewery started in 1896.[420]

Siebel Institute of Technology, a center for brewing education since 1872, moved from northern Chicago to 1700 N. Clybourn St. at the Goose Island brewpub.[421] Siebel had announced its intention to close in 2000.

Reduced calorie beers made up 45 percent of the U.S. market.[422]

2004 The fad low-carb diet led to new low-carb beers. Overall beer consumption dropped in response to the diet's requirements.[423]

Forty percent of beer was canned, 43 percent was bottled, and 8 percent was served on tap. Only 0.5 percent was sold in plastic bottles.[424]

The aluminum beer bottle was introduced in the United States.[425] It was shaped like a bottle and capped, and all aluminum.

Adolph Coors Co. and Canada's Molson announced a merger that would create a brewer with $6 billion in sales.

Anheuser-Busch bought Harbin Brewery Group to expand its presence in China.[426]

2007 SABMiller Brewing and Molson Coors announced a merger of their domestic operations.

The problem of keg theft worsened; brewers called for junk dealers to demand proof of ownership from individuals trying to sell kegs.[427]

Michael Jackson, the Beer Hunter, passed away. His videos, books, and magazine articles changed beer consumption around the world.[428]

2008 SABMiller PLC and Molson Coors Brewing Co. closed on their transaction, combining their U.S. and Puerto Rico operations into MillerCoors.

Craft breweries as a group sold over 8.5 million barrels in 2008.

2009 National Brewing Museum opened in Potosi, Wisconsin.

Craft breweries as a group sold approximately 9.12 million barrels in 2009.

2010 The ten largest U.S. craft breweries accounted for over 4.5 million barrels annually.

C. Dean Metropoulos bought Pabst Brewing Company, owner of the Schlitz, Old Style, Lone Star, and Colt 45 brands. [429] The parent company owns no brewery and instead contracts for its beer.

The European Court of Justice rejected the trademark registration request from Anheuser-Busch InBev for European Union-wide trademark status. This was the final blow in the decades-long effort by Anheuser-Busch InBev to gain control of the Budweiser name in Europe.[430]

NOTES

1. Charles Gibson, *The Aztecs Under Spanish Rule: A History of the Indians of the Valley of Mexico* (Palo Alto, California: Stanford University Press, 1976), 347–8.
2. Bull, Friedrich, and Gottschalk, *American Breweries II*, 1.
3. Bull, Friedrich, and Gottschalk, *American Breweries II*, 1.
4. Allen Winn Sneath, *Brewed in Canada: The Untold Story of Canada's 350-Year-Old Brewing Industry* (Toronto: The Dundurn Group, 2001), 20.
5. Brown, *Wines and Beers of Old New England: A How-to-Do-It History*, 63.
6. Bull, Friedrich, and Gottschalk, *American Breweries II*, 1.
7. George Ehret, *Twenty-Five Years of Brewing With an Illustrated History of American Beer* (New York: George Ehret Brewery, 1891), 7.
8. Bull, Friedrich, and Gottschalk, *American Breweries II*, 1.
9. Ehret, *Twenty-Five Years of Brewing With an Illustrated History of American Beer*, 11.
10. Bull, Friedrich, and Gottschalk, *American Breweries II*, 1.
11. Gregg Smith, *A History of Suds And Civilization From Mesopotamia to Microbreweries* (New York: Avon Books, 1995), 37.
12. Tomlan, *Tinged with Gold: Hop Culture in the United States*, 11.
13. Sneath, *Brewed in Canada: The Untold Story of Canada's 350-Year-Old Brewing Industry*, 326.
14. Ehret, *Twenty-Five Years of Brewing With an Illustrated History of American Beer*, 26.
15. *100 Years of Brewing* (1903; repr. New York: Arno Press, 1974), 174.
16. Arnold and Penman, *History of the Brewing Industry and Brewing Science in America*, 48.
17. Brown, *Wines and Beers of Old New England: A How-to-Do-It History*, 52.
18. Smith, *A History of Suds And Civilization From Mesopotamia to Microbreweries*, 33.
19. Bull, Friedrich, and Gottschalk, *American Breweries II*, 1.
20. Bull, Friedrich, and Gottschalk, *American Breweries II*, 1.
21. Apps, *Breweries of Wisconsin*, 9.
22. Anderson, *Beer, USA*, 3.
23. Bull, Friedrich, and Gottschalk, *American Breweries II*, 1.
24. Smith, *A History of Suds And Civilization From Mesopotamia to Microbreweries*, 82.
25. Polak, *Bottles: Identification and Price Guide*, 2nd ed., 7.
26. Appel, "The Midwestern Brewery Before Prohibition: Development of an American Industrial Building Type," 58.
27. Smith, *A History of Suds And Civilization From Mesopotamia to Microbreweries*, 63.
28. *100 Years of Brewing*, 1974 reprint, 173–4.
29. Bull, Friedrich, and Gottschalk, *American Breweries II*, 1.
30. Bull, Friedrich, and Gottschalk, *American Breweries II*, 1.
31. The richest inhabitant of Kaskaskia in1766 was Jean Baptiste St. Gerome Beauvis. See *Combined History of Randolph, Monroe and Perry Counties, Illinois* (Philadelphia: J. L. McDonough & Co., 1883), 63.
32. Bull, Friedrich, and Gottschalk, *American Breweries II*, 1.
33. Smith, *A History of Suds And Civilization From Mesopotamia to Microbreweries*, 49.
34. Bill Yenn, *The American Brewery: From Colonial Evolution to Microbrew Revolution* (St. Paul, MN: MBI, 2003), 15.
35. Ehret, *Twenty-Five Years of Brewing With an Illustrated History of American Beer*, 31.
36. W. J. Rorabaugh, *The Alcoholic Republic: An American Tradition* (Oxford: Oxford University Press, 1979), 65.

37. Rorabaugh, *The Alcoholic Republic: An American Tradition,* 45.
38. Guetig and Selle, *Louisville Breweries: A History of the Brewing Industry in Louisville, Kentucky, New Albany and Jeffersonville, Indiana,* 12.
39. Amy Mittelman, "The Politics of Alcohol Production: The Liquor Industry and the Federal Government, 1862–1900" (dissertation, Columbia University, 1986), 150.
40. Smith, *A History of Suds And Civilization From Mesopotamia to Microbreweries,* 96.
41. Tomlan, *Tinged with Gold: Hop Culture in the United States,* 73.
42. *100 Years of Brewing,* 1974 reprint, 88.
43. Arnold and Penman, *History of the Brewing Industry and Brewing Science in America,* Chicago, 1933, p. 92.
44. *100 Years of Brewing,* 1974 reprint, 78.
45. *100 Years of Brewing,* 1974 reprint, 135.
46. *100 Years of Brewing,* 1974 reprint, 82–83.
47. Bernard Devoto, ed., *The Journals of Lewis and Clark* (Boston: Houghton Mifflin, 1953), 260. This was on the Columbia River above Celilo Falls.
48. Tomlan, *Tinged with Gold: Hop Culture in the United States,* 13.
49. Appel, "The Midwestern Brewery Before Prohibition: Development of an American Industrial Building Type," 26.
50. Appel, "The Midwestern Brewery Before Prohibition: Development of an American Industrial Building Type," 61.
51. Ronnenberg, *Beer and Brewing in the Inland Northwest,* 84.
52. Appel, "The Midwestern Brewery Before Prohibition: Development of an American Industrial Building Type," 68.
53. Powell, *Rushing the Growler: A History of Brewing in Buffalo,* 5.
54. Arnold and Penman, *History of the Brewing Industry and Brewing Science in America,* Chicago, 1933, p. 90.
55. Appel, "The Midwestern Brewery Before Prohibition: Development of an American Industrial Building Type," 66.
56. Appel, "The Midwestern Brewery Before Prohibition: Development of an American Industrial Building Type," 65.
57. *100 Years of Brewing,* 1974 reprint, 195. This was probably the second in Indiana and the third in what was considered the West of that day.
58. Appel, "The Midwestern Brewery Before Prohibition: Development of an American Industrial Building Type," 59.
59. Appel, "The Midwestern Brewery Before Prohibition: Development of an American Industrial Building Type," 68.
60. Arnold and Penman, *History of the Brewing Industry and Brewing Science in America,* p. 54.
61. *100 Years of Brewing,* 1974 reprint, 78.
62. Peter H. Blum, *Brewed in Detroit: Breweries and Beers Since 1830* (Detroit: Wayne State University Press, 1999), 41. Miller, *Breweries of Cleveland,* 19.
63. Miller, *Breweries of Cleveland,* 21.
64. Tomlan, *Tinged with Gold: Hop Culture in the United States,* 80.
65. Apps, *Breweries of Wisconsin,* 14.
66. Smith, *A History of Suds And Civilization From Mesopotamia to Microbreweries,* 95.
67. Bull, Friedrich, and Gottschalk, *American Breweries II,* 2.
68. Smith, *A History of Suds And Civilization From Mesopotamia to Microbreweries,* 153.
69. Downard, *The Cincinnati Brewing Industry, A Social and Economic History,* 11.
70. Appel, "The Midwestern Brewery Before Prohibition: Development of an American Industrial-Building Type," 97.
71. Mittelman, "The Politics of Alcohol Production: The Liquor Industry and the Federal Government, 1862–1900," 151.
72. Tomlan, *Tinged with Gold: Hop Culture in the United States,* 13.
73. Appel, "The Midwestern Brewery Before Prohibition: Development of an American Industrial Building Type," 73.
74. *100 Years of Brewing,* 1974 reprint, 61.
75. Scammel and Colyer, *Breweries and Maltings: Their Arrangement, Construction, Machinery, and Plant,* 102.
76. *100 Years of Brewing,* 1974 reprint, 148.
77. Ronnenberg, *Beer and Brewing in the Inland Northwest,* 13.
78. Rother & Rother, *Lost Caves of St. Louis: A History of the City's Forgotten Caves,* 25.
79. Bull, Friedrich, and Gottschalk, *American Breweries II,* 3.
80. Ronnenberg, "The American Brewing Industry Since 1920," *The Dynamics of the International Brewing Industry Since 1800* (New York: Routledge, 1998), 209.
81. Dick, "A Geographical Analysis of the Development of the Brewing Industry of Minnesota," 81, 182.
82. Herbert B. Nelson, "The Vanishing Hop-Driers of the Willamette Valley," *Oregon Historical Quarterly* 64 no. 3 (1963), 267.
83. Bull, Friedrich, and Gottschalk, *American Breweries II,* 3.
84. Appel, "The Midwestern Brewery Before Prohibition: Development of an American Industrial Building Type," 88.
85. *100 Years of Brewing,* 1974 reprint, 84.
86. Bull, Friedrich, and Gottschalk, *American Breweries II,* 3.
87. *100 Years of Brewing,* 1974 reprint, 136.
88. Ronnenberg, *Beer and Brewing in the Inland Northwest,* 13.
89. Bull, Friedrich, and Gottschalk, *American Breweries II,* 3.
90. Smith, *A History of Suds And Civilization From Mesopotamia to Microbreweries,* 146.
91. Ronnenberg, *Beer and Brewing in the Inland Northwest,* 14.
92. Guetig, and Selle, *Louisville Breweries: A History of the Brewing Industry in Louisville, Kentucky, New Albany and Jeffersonville, Indiana,* 16.

93. Ronnenberg, *Beer and Brewing in the Inland Northwest*, 12.
94. Miller, *Breweries of Cleveland*, 34.
95. Tomlan, *Tinged with Gold: Hop Culture in the United States*, 27.
96. Morris and Johnson, *Richmond Beers, A Directory of the Breweries and Bottles of Richmond, Virginia*, 2nd ed., 100.
97. Smith, *A History of Suds And Civilization From Mesopotamia to Microbreweries*, 99.
98. *100 Years of Brewing*, 1974 reprint, 359.
99. Mittelman, "The Politics of Alcohol Production: The Liquor Industry and the Federal Government, 1862–1900." 196.
100. Arnold and Penman, *History of the Brewing Industry and Brewing Science in America*, 97.
101. Ronnenberg, *Beer and Brewing in the Inland Northwest*, 90.
102. Millard, Arnold A. *History of the Carson Brewing Company*, p. 7.
103. *100 Years of Brewing*, 1974 reprint, 364.
104. Downard, *The Cincinnati Brewing Industry, A Social and Economic History*, 30.
105. Mittelman, "The Politics of Alcohol Production: The Liquor Industry and the Federal Government, 1862–1900," 152.
106. Bull, Friedrich, and Gottschalk, *American Breweries II*, 3.
107. *100 Years of Brewing*, 1974 reprint, 125.
108. Bull, Friedrich, and Gottschalk, *American Breweries II*, 4.
109. Appel, "The Midwestern Brewery Before Prohibition: Development of an American Industrial Building Type," 118.
110. *100 Years of Brewing*, 1974 reprint, 117.
111. Smith, *A History of Suds And Civilization From Mesopotamia to Microbreweries*, 102.
112. Tomlan, *Tinged with Gold: Hop Culture in the United States*, 192–193.
113. *100 Years of Brewing*, 1974 reprint, 84.
114. Tomlan, *Tinged with Gold: Hop Culture in the United States*, 80.
115. Mittelman, "The Politics of Alcohol Production: The Liquor Industry and the Federal Government, 1862–1900," 167.
116. *100 Years of Brewing*, 1974 reprint, 63.
117. *100 Years of Brewing*, 1974 reprint, 125.
118. *100 Years of Brewing*, 1974 reprint, 136.
119. Ehret, *Twenty-Five Years of Brewing With an Illustrated History of American Beer*, 47.
120. Ogle, *Ambitious Brew*, 57.
121. *100 Years of Brewing*, 1974 reprint, 84.
122. Mittelman, "The Politics of Alcohol Production: The Liquor Industry and the Federal Government, 1862–1900," 176.
123. *100 Years of Brewing*, 1974 reprint, 151.
124. Ronnenberg, *Beer and Brewing in the Inland Northwest*, 39.
125. Tomlan, *Tinged with Gold: Hop Culture in the United States*, 193.
126. Arnold and Penman, *History of the Brewing Industry and Brewing Science in America*, 93.
127. Tomlan, *Tinged with Gold: Hop Culture in the United States*, 197.
128. Cooler, *Hop Agriculture in Oregon: The First Century*, 21.
129. Ogle, *Ambitious Brew*, 57.
130. Arnold and Penman, *History of the Brewing Industry and Brewing Science in America*, 93.
131. *100 Years of Brewing*, 1974 reprint, 125.
132. *100 Years of Brewing*, 1974 reprint, 135.
133. *100 Years of Brewing*, 1974 reprint, 57.
134. *100 Years of Brewing*, 1974 reprint, 120.
135. Bull, Friedrich, and Gottschalk, *American Breweries II*, 4.
136. Skilnik, *The History of Beer and Brewing in Chicago, 1833–1978*, 22–23.
137. Mittelman, "The Politics of Alcohol Production: The Liquor Industry and the Federal Government, 1862–1900," 177.
138. Mittelman, "The Politics of Alcohol Production: The Liquor Industry and the Federal Government, 1862–1900," 200.
139. Bull, Friedrich, and Gottschalk, *American Breweries II*, 4. Appel, "The Midwestern Brewery Before Prohibition: Development of an American Industrial Building Type," 40.
140. K. Austin Kerr, "The American Brewing Industry, 1865–1920," *The Dynamics of the International Brewing Industry since 1800* (New York: Routledge, 1998), 184.
141. Appel, "The Midwestern Brewery Before Prohibition: Development of an American Industrial Building Type," 44.
142. Cochran, *The Pabst Brewing Company: The History of an American Business*, 118.
143. Ogle, *Ambitious Brew*, 64.
144. *100 Years of Brewing*, 1974 reprint, 125.
145. *100 Years of Brewing*, 1974 reprint, 63.
146. Morris and Johnson, *Richmond Beers, A Directory of the Breweries and Bottles of Richmond, Virginia*, 2nd ed., 2000, 100.
147. Tomlan, *Tinged with Gold: Hop Culture in the United States*, 28.
148. Appel, "The Midwestern Brewery Before Prohibition: Development of an American Industrial Building Type," 38.
149. *100 Years of Brewing*, 1974 reprint, 57.
150. *100 Years of Brewing*, 1974 reprint, 124.
151. *100 Years of Brewing*, 1974 reprint, 129.
152. *100 Years of Brewing*, 1974 reprint, 149.
153. *100 Years of Brewing*, 1974 reprint, 136–7.

154. Smith, *A History of Suds And Civilization From Mesopotamia to Microbreweries*, 134.
155. *100 Years of Brewing*, 1974 reprint, 125.
156. *100 Years of Brewing*, 1974 reprint, 126.
157. *100 Years of Brewing*, 1974 reprint, 128.
158. *100 Years of Brewing*, 1974 reprint, 130.
159. Appel, "The Midwestern Brewery Before Prohibition: Development of an American Industrial Building Type," 225.
160. Stack, "Liquid Bread: An Examination of the American Brewing Industry, 1865–1940," 85.
161. *100 Years of Brewing*, 1974 reprint, 127.
162. *100 Years of Brewing*, 1974 reprint, 134.
163. Appel, "The Midwestern Brewery Before Prohibition: Development of an American Industrial Building Type," note 200, p. 255.
164. Cochran, *The Pabst Brewing Company: The History of an American Business*, 111.
165. *100 Years of Brewing*, 1974 reprint, 126.
166. *100 Years of Brewing*, 1974 reprint, 131.
167. Ronnenberg, *Beer and Brewing in the Inland Northwest*, 40.
168. Cochran, *The Pabst Brewing Company: The History of an American Business*, 103.
169. *The Idaho Semi-Weekly World* (Idaho City, Idaho), June 17, 1879, p. 3, c. 3.
170. Appel, "The Midwestern Brewery Before Prohibition: Development of an American Industrial Building Type," 44.
171. Arnold and Penman, *History of the Brewing Industry and Brewing Science in America*, 106.
172. Cochran, *The Pabst Brewing Company: The History of an American Business*, 423.
173. Tomlan, *Tinged with Gold: Hop Culture in the United States*, 22.
174. *100 Years of Brewing*, 1974 reprint, 152.
175. *100 Years of Brewing*, 1974 reprint, 122.
176. *100 Years of Brewing*, 1974 reprint, 132–133.
177. *100 Years of Brewing*, 1974 reprint, 136.
178. *100 Years of Brewing*, 1974 reprint, 137.
179. *100 Years of Brewing*, 1974 reprint, 63.
180. *100 Years of Brewing*, 1974 reprint, 73.
181. Mittelman, "The Politics of Alcohol Production: The Liquor Industry and the Federal Government, 1862–1900," 201.
182. *100 Years of Brewing*, 1974 reprint, 64.
183. Appel, "The Midwestern Brewery Before Prohibition: Development of an American Industrial Building Type," 202.
184. *100 Years of Brewing*, 1974 reprint, 130.
185. *100 Years of Brewing*, 1974 reprint, 134.
186. Appel, "The Midwestern Brewery Before Prohibition: Development of an American Industrial Building Type," 210.
187. *100 Years of Brewing*, 1974 reprint, 85.
188. Cochran, *The Pabst Brewing Company: The History of an American Business*, 125.
189. Mittelman, "The Politics of Alcohol Production: The Liquor Industry and the Federal Government, 1862–1900," 202–203.
190. Ronnenberg, *Beer and Brewing in the Inland Northwest*, 73.
191. Arnold and Penman, *History of the Brewing Industry and Brewing Science in America*, 96.
192. Bull, Friedrich, and Gottschalk, *American Breweries II*, 5.
193. *100 Years of Brewing*, 1974 reprint, 133.
194. *100 Years of Brewing*, 1974 reprint, 151.
195. Mittelman, "The Politics of Alcohol Production: The Liquor Industry and the Federal Government, 1862–1900," 203.
196. Mittelman, "The Politics of Alcohol Production: The Liquor Industry and the Federal Government, 1862–1900," 206.
197. Peter H. Blum, *Brewed in Detroit: Breweries and Beers Since 1830* (Detroit: Wayne State University Press, 1999), 125.
198. Skilnik, *The History of Beer and Brewing in Chicago, 1833–1978*, 51.
199. Blum, *Brewed in Detroit: Breweries and Beers Since 1830*, 37.
200. *100 Years of Brewing*, 1974 reprint, 134.
201. Cochran, *The Pabst Brewing Company: The History of an American Business*, 106.
202. Cochran, *The Pabst Brewing Company: The History of an American Business*, 112.
203. *100 Years of Brewing*, 1974 reprint, 120.
204. *U.S. Bureau of the Census*, Report of the Eleventh Census, 1890; *100 Years of Brewing*, 1974 reprint, 604.
205. *100 Years of Brewing*, 1974 reprint, 145.
206. Cochran, *The Pabst Brewing Company: The History of an American Business*, 109.
207. Mittelman, "The Politics of Alcohol Production: The Liquor Industry and the Federal Government, 1862–1900," 222.
208. Cochran, *The Pabst Brewing Company: The History of an American Business*, 106.
209. Cochran, *The Pabst Brewing Company: The History of an American Business*, 110.
210. Cochran, *The Pabst Brewing Company: The History of an American Business*, 107.
211. Cochran, *The Pabst Brewing Company: The History of an American Business*, 126–127.
212. *100 Years of Brewing*, 1974 reprint, 125.

213. Appel, "The Midwestern Brewery Before Prohibition: Development of an American Industrial Building Type," 216. Cochran, *The Pabst Brewing Company: The History of an American Business,* 105.
214. *Idaho Daily Statesman* (Boise, Idaho), September 2, 1891, p. 8, c. 1.
215. Bull, Friedrich, and Gottschalk, *American Breweries II,* 1995, 5.
216. Bull, Friedrich, and Gottschalk, *American Breweries II,* 1995, 5.
217. Kerr, "The American Brewing Industry, 1865–1920," 184.
218. Kerr, "The American Brewing Industry, 1865–1920," 185.
219. *100 Years of Brewing,* 1974 reprint, 127.
220. Cochran, *The Pabst Brewing Company: The History of an American Business,* 123.
221. Ogle, *Ambitious Brew,* 2006, 125.
222. *100 Years of Brewing,* 1974 reprint, 66.
223. Stack, "Liquid Bread: An Examination of the American Brewing Industry, 1865–1940," 132.
224. *100 Years of Brewing,* 1974 reprint, 71.
225. *100 Years of Brewing,* 1974 reprint, 71.
226. Stack, "Liquid Bread: An Examination of the American Brewing Industry, 1865–1940," 116.
227. Mittelman, "The Politics of Alcohol Production: The Liquor Industry and the Federal Government, 1862–1900," 223.
228. Appel, "The Midwestern Brewery Before Prohibition: Development of an American Industrial Building Type," 39.
229. Mittelman, "The Politics of Alcohol Production: The Liquor Industry and the Federal Government, 1862–1900," 184.
230. Cooler, *Hop Agriculture in Oregon: The First Century,* 22.
231. *U.S. Bureau of the Census,* Report of the Eleventh Census, 1890; *100 Years of Brewing,* 1974 reprint, 604.
232. Anderson, *Beer, USA,* 17.
233. Cooler, *Hop Agriculture in Oregon: The First Century,* 36–37.
234. Mittelman, "The Politics of Alcohol Production: The Liquor Industry and the Federal Government, 1862–1900," 225–226.
235. Stack, "Liquid Bread: An Examination of the American Brewing Industry, 1865–1940," 131.
236. Mittelman, "The Politics of Alcohol Production: The Liquor Industry and the Federal Government, 1862–1900," 188.
237. *Idaho Daily Statesman* (Boise, Idaho), June 30, 1900.
238. Anderson, *Beer, USA,* 18.
239. Downard, *The Cincinnati Brewing Industry, A Social and Economic History,* 114–120.
240. Polak, *Bottles: Identification and Price Guide,* 2nd ed., 11.
241. Stack, "Liquid Bread: An Examination of the American Brewing Industry, 1865–1940," 117.
242. Appel, "The Midwestern Brewery Before Prohibition: Development of an American Industrial Building Type," 32.
243. Harris, "From Limestone to Rubble at a Brewery Full of Memories," 17.
244. *The Western Brewer* vol. XXXIII, no. 3 (March 15, 1908), 126. *Idaho Register* (Idaho Falls), October 20, 1908, p. 3, c. 3.
245. Gregg, Page, and Waggoner, eds., *United States Beer Cans: The Standard Reference of Flat Tops and Cone Tops,* 6.
246. Tomlan, *Tinged with Gold: Hop Culture in the United States,* 114.
247. Ronnenberg, "The Sunset Brewery of Wallace, Idaho," *The Breweriana Collector* 38 (Summer 1982), 19.
248. Ronnenberg, *Beer and Brewing in the Inland Northwest,* 81.
249. Anderson, *Beer, USA,* 24.
250. Anderson, *Beer, USA,* 25.
251. Anderson, *Beer, USA,* 26.
252. Anderson, *Beer, USA,* 28.
253. Dick, "A Geographical Analysis of the Development of the Brewing Industry of Minnesota," 59.
254. Ronnenberg, *The Politics of Assimilation,* 76.
255. Ronnenberg, *The Politics of Assimilation,* 76.
256. Ronnenberg, *The Politics of Assimilation,* 76.
257. Ronnenberg, "The American Brewing Industry Since 1920," 193.
258. Tomlan, *Tinged with Gold: Hop Culture in the United States,* 38.
259. Miller, *Breweries of Cleveland,* 147.
260. Bull, Friedrich, and Gottschalk, *American Breweries II,* 6.
261. Smith, *A History of Suds And Civilization From Mesopotamia to Microbreweries,* 118.
262. Ronnenberg, *The Politics of Assimilation,* 79.
263. Rother & Rother, *Lost Caves of St. Louis: A History of the City's Forgotten Caves,* 30.
264. Ronnenberg, *The Politics of Assimilation,* 82.
265. Skilnik, *The History of Beer and Brewing in Chicago, 1833–1978,* 123.
266. Ronnenberg, *The Politics of Assimilation,* 84.
267. Yenn, *The American Brewery: From Colonial Evolution to Microbrew Revolution,* 77.
268. Smith, *A History of Suds And Civilization From Mesopotamia to Microbreweries,* 118.
269. Cooler, *Hop Agriculture in Oregon: The First Century,* 48.
270. Ronnenberg, *The Politics of Assimilation,* 98.
271. Ronnenberg, *The Politics of Assimilation,* 82.
272. Lilek, *United States Beer Cans with Opening Instructions, The Standard Reference Guide of Opening Instruction Flat Tops,* iv.

273. Ronnenberg, "The American Brewing Industry Since 1920," 198.
274. Lilek, *United States Beer Cans with Opening Instructions, The Standard Reference Guide of Opening Instruction Flat Tops,* iv.
275. Rother & Rother, *Lost Caves of St. Louis: A History of the City's Forgotten Caves,* 57.
276. Dick, "A Geographical Analysis of the Development of the Brewing Industry of Minnesota," 65.
277. Anderson, *Beer, USA,* 51.
278. Anderson, *Beer, USA,* 52.
279. Anderson, *Beer, USA,* 141.
280. Larry Wright, ed., *The Beer Can: A Complete Guide to Beer Can Collecting* (Matteson, Illinois: Great Lakes Living Press, 1976), 3.
281. Bull, Friedrich, and Gottschalk, *American Breweries II,* 7.
282. Maxwell, "Beer Cans: A Guide for the Archaeologist," 96.
283. Bull, Friedrich, and Gottschalk, *American Breweries II,* p. 7.
284. Anderson, *Beer, USA,* 53.
285. Anderson, *Beer, USA,* 54.
286. Gregg, Page, and Waggoner, eds., *United States Beer Cans: The Standard Reference of Flat Tops and Cone Tops,* 10.
287. Cooler, *Hop Agriculture in Oregon: The First Century,* 39.
288. Morris Weeks, Jr., *Beer and Brewing in America,* (New York: United States Brewers Foundation, 1949), 71.
289. Ronnenberg, "The American Brewing Industry Since 1920," 203.
290. Dick, "A Geographical Analysis of the Development of the Brewing Industry of Minnesota," 65.
291. Maxwell, "Beer Cans: A Guide for the Archaeologist," 96.
292. Gregg, Page, and Waggoner, eds., *United States Beer Cans: The Standard Reference of Flat Tops and Cone Tops,* 13.
293. Lilek, *United States Beer Cans with Opening Instructions, The Standard Reference Guide of Opening Instruction Flat Tops,* p. v.
294. Tomlan, *Tinged with Gold: Hop Culture in the United States,* 114.
295. Tomlan, *Tinged with Gold: Hop Culture in the United States,* 33.
296. Ronnenberg, "The American Brewing Industry Since 1920," 206.
297. Smith, *A History of Suds And Civilization From Mesopotamia to Microbreweries,* 99.
298. Maxwell, "Beer Cans: A Guide for the Archaeologist," 96.
299. Anderson, *Beer, USA,* 6.
300. Anderson, *Beer, USA,* 141.
301. Dick, "A Geographical Analysis of the Development of the Brewing Industry of Minnesota," 66.
302. Gregg, Page, and Waggoner, eds., *United States Beer Cans: The Standard Reference of Flat Tops and Cone Tops,* 12.
303. Anderson, *Beer, USA,* 79.
304. Martells, *The Beer Can Collectors Bible,* 9.
305. Bull, Friedrich, and Gottschalk, *American Breweries II,* 7.
306. Maxwell, "Beer Cans: A Guide for the Archaeologist," 96.
307. Anderson, *Beer, USA,* 83.
308. Anderson, *Beer, USA,* 122.
309. Weeks, *Beer and Brewing in America,* 71–72.
310. Anderson, *Beer, USA,* 84.
311. Cooler, *Hop Agriculture in Oregon: The First Century,* p. 67.
312. Anderson, *Beer, USA,* 126.
313. Anderson, *Beer, USA,* 141.
314. Weeks, *Beer and Brewing in America,* 28–29.
315. Anderson, *Beer, USA,* 127.
316. Anderson, *Beer, USA,* 129.
317. Anderson, *Beer, USA,* 134.
318. Anderson, *Beer, USA,* 135.
319. Anderson, *Beer, USA,* 138.
320. Tomlan, *Tinged with Gold: Hop Culture in the United States,* 115.
321. Maxwell, "Beer Cans: A Guide for the Archaeologist," 96.
322. Kroll, *Badger Breweries Past and Present,* 7.
323. Anderson, *Beer, USA,* 141.
324. Yenn, *The American Brewery: From Colonial Evolution to Microbrew Revolution,* 97.
325. Bull, Friedrich, and Gottschalk, *American Breweries II,* 7.
326. Anderson, *Beer, USA,* 141.
327. Martells, *The Beer Can Collectors Bible,* 9.
328. Cooler, *Hop Agriculture in Oregon: The First Century,* 3.
329. Dick, "A Geographical Analysis of the Development of the Brewing Industry of Minnesota," 68,149.
330. Anderson, *Beer, USA,* 143.
331. Anderson, *Beer, USA,* 145.
332. Anderson, *Beer, USA,* 145.
333. Maxwell, "Beer Cans: A Guide for the Archaeologist," 96.
334. Maxwell, "Beer Cans: A Guide for the Archaeologist," 96.
335. Blum, *Brewed in Detroit: Breweries and Beers Since 1830,* 77.
336. Bull, Friedrich, and Gottschalk, *American Breweries II,* 7.
337. Anderson, *Beer, USA,* 154.
338. Gregg, Page, and Waggoner, eds., *United States Beer Cans: The Standard Reference of Flat Tops and Cone Tops,* 12.
339. Bull, Friedrich, and Gottschalk, *American Breweries II,* 7.
340. Anderson, *Beer, USA,* 154.
341. Gregg, Page, and Waggoner, eds., *United States Beer Cans: The Standard Reference of Flat Tops and Cone Tops,* 14.
342. Yenn, *The American Brewery: From Colonial Evolution to Microbrew Revolution,* 97.
343. Bull, Friedrich, and Gottschalk, *American Breweries II,* 7.

344. John Vetter, "What is Unique About These Crowns?" *American Breweriana Journal* 144 (November-December, 2006), 36.
345. Maxwell, "Beer Cans: A Guide for the Archaeologist," 96.
346. Maxwell, "Beer Cans: A Guide for the Archaeologist," 96.
347. Tomlan, *Tinged with Gold: Hop Culture in the United States,* 41.
348. Anderson, *Beer, USA,* 154.
349. Anderson, *Beer, USA,* 154.
350. Maxwell, "Beer Cans: A Guide for the Archaeologist," 96.
351. Bull, Friedrich, and Gottschalk, *American Breweries II,* 8.
352. Anderson, *Beer, USA,* 155.
353. Martells, *The Beer Can Collectors Bible,* 10.
354. Maxwell, "Beer Cans: A Guide for the Archaeologist," 96.
355. Appel, "The Midwestern Brewery Before Prohibition: Development of an American Industrial Building Type," 346.
356. Maxwell, "Beer Cans: A Guide for the Archaeologist," 96.
357. Apps, *Breweries of Wisconsin,* 108.
358. Bull, Friedrich, and Gottschalk, *American Breweries II,* 8.
359. Bull, Friedrich, and Gottschalk, *American Breweries II,* 8.
360. Maxwell, "Beer Cans: A Guide for the Archaeologist," 96.
361. Maxwell, "Beer Cans: A Guide for the Archaeologist," 96.
362. Wright, *The Beer Can: A Complete Guide to Beer Can Collecting,* 24.
363. Anderson, *Beer, USA,* 156.
364. Maxwell, "Beer Cans: A Guide for the Archaeologist," 109.
365. Yenn, *The American Brewery: From Colonial Evolution to Microbrew Revolution,* 97.
366. Ronnenberg, "The American Brewing Industry Since 1920," 208.
367. Dick, "A Geographical Analysis of the Development of the Brewing Industry of Minnesota," 292.
368. Maxwell, "Beer Cans: A Guide for the Archaeologist," 96.
369. Maxwell, "Beer Cans: A Guide for the Archaeologist," 96.
370. *Real Beer Page Mail, The Free Monthly Beer News Digest for the Online Beer Enthusiast,* rbpmail@realbeer.com, January 11, 2003.
371. Anderson, *Beer, USA,* 158.
372. *Real Beer Page Mail, The Free Monthly Beer News Digest for the Online Beer Enthusiast,* rbpmail@realbeer.com, January 11, 2003.
373. Apps, *Breweries of Wisconsin,* 109.
374. Bull, Friedrich, and Gottschalk, *American Breweries II,* 8.
375. Maxwell, "Beer Cans: A Guide for the Archaeologist," 96.
376. Anderson, *Beer, USA,* 161.
377. Maxwell, "Beer Cans: A Guide for the Archaeologist," 96.
378. Mittelman, "The Politics of Alcohol Production: The Liquor Industry and the Federal Government, 1862–1900," 150.
379. Yenn, *The American Brewery: From Colonial Evolution to Microbrew Revolution,* 97.
380. Appel, "The Midwestern Brewery Before Prohibition: Development of an American Industrial Building Type," 346.
381. Apps, *Breweries of Wisconsin,* 111.
382. Bull, Friedrich, and Gottschalk, *American Breweries II,* 8.
383. Bull, Friedrich, and Gottschalk, *American Breweries II,* 8.
384. Maxwell, "Beer Cans: A Guide for the Archaeologist," 96.
385. Bull, Friedrich, and Gottschalk, *American Breweries II,* 8.
386. Maxwell, "Beer Cans: A Guide for the Archaeologist," 96.
387. *Real Beer Page Mail, The Free Monthly Beer News Digest for the Online Beer Enthusiast,* rbpmail@realbeer.com, November 1998.
388. Blum, *Brewed in Detroit: Breweries and Beers Since 1830,* 86.
389. Apps, *Breweries of Wisconsin,* 136.
390. Maxwell, "Beer Cans: A Guide for the Archaeologist," 96.
391. Yenn, *The American Brewery: From Colonial Evolution to Microbrew Revolution,* 97.
392. Bull, Friedrich, and Gottschalk, *American Breweries II,* 8.
393. Ogle, *Ambitious Brew,* 331.
394. Christopher B. O'Hara, *Great American Beer: 50 Brands that Shaped the 20th Century* (New York: Clarkson Potter, 2006), 36.
395. O'Hara, *Great American Beer, 50 Brands that Shaped the 20th Century,* 37.
396. *Real Beer Page Mail, The Free Monthly Beer News Digest for the Online Beer Enthusiast,* rbpmail@realbeer.com, November 1998.
397. O'Hara, *Great American Beer, 50 Brands that Shaped the 20th Century,* 37.
398. *Real Beer Page Mail, The Free Monthly Beer News Digest for the Online Beer Enthusiast,* rbpmail@realbeer.com, July 1999.
399. *Real Beer Page Mail, The Free Monthly Beer News Digest for the Online Beer Enthusiast,* rbpmail@realbeer.com, October 1998.
400. *Real Beer Page Mail, The Free Monthly Beer News Digest for the Online Beer Enthusiast,* rbpmail@realbeer.com, November 1998.
401. *Real Beer Page Mail, The Free Monthly Beer News Digest for the Online Beer Enthusiast,* rbpmail@realbeer.com, November 1998.
402. *Real Beer Page Mail, The Free Monthly Beer News Digest for the Online Beer Enthusiast,* rbpmail@realbeer.com, May 1999.

403. *Real Beer Page Mail, The Free Monthly Beer News Digest for the Online Beer Enthusiast,* rbpmail@realbeer.com, October 2000.
404. *Real Beer Page Mail, The Free Monthly Beer News Digest for the Online Beer Enthusiast,* rbpmail@realbeer.com, March 2000.
405. *Real Beer Page Mail, The Free Monthly Beer News Digest for the Online Beer Enthusiast,* rbpmail@realbeer.com, April 2000.
406. Yenn, *The American Brewery: From Colonial Evolution to Microbrew Revolution,* 143, 146.
407. O'Hara, *Great American Beer, 50 Brands that Shaped the 20th Century,* 37.
408. O'Hara, *Great American Beer, 50 Brands that Shaped the 20th Century,* 37.
409. *Real Beer Page Mail, The Free Monthly Beer News Digest for the Online Beer Enthusiast,* rbpmail@realbeer.com, October 2001.
410. *Real Beer Page Mail, The Free Monthly Beer News Digest for the Online Beer Enthusiast,* rbpmail@realbeer.com, March 2001.
411. *Real Beer Page Mail, The Free Monthly Beer News Digest for the Online Beer Enthusiast,* rbpmail@realbeer.com, September 2001.
412. O'Hara, *Great American Beer, 50 Brands that Shaped the 20th Century,* 37.
413. Ogle, *Ambitious Brew,* 336.
414. "Logistics Firm Offers Kegs for Lease," *American Breweriana Journal* 118 (July-August 2002), 37.
415. *Real Beer Page Mail, The Free Monthly Beer News Digest for the Online Beer Enthusiast,* rbpmail@realbeer.com, July 2002, 1.
416. *Real Beer Page Mail, The Free Monthly Beer News Digest for the Online Beer Enthusiast,* rbpmail@realbeer.com, March 2002, 1–2.
417. *Real Beer Page Mail, The Free Monthly Beer News Digest for the Online Beer Enthusiast,* rbpmail@realbeer.com, March 2002, 8.
418. O'Hara, *Great American Beer, 50 Brands that Shaped the 20th Century,* 37.
419. *Real Beer Page Mail, The Free Monthly Beer News Digest for the Online Beer Enthusiast,* rbpmail@realbeer.com, September 2003.
420. *Real Beer Page Mail, The Free Monthly Beer News Digest for the Online Beer Enthusiast,* January 2003.
421. *Real Beer Page Mail, The Free Monthly Beer News Digest for the Online Beer Enthusiast,* rbpmail@realbeer.com, February 2003.
422. *Real Beer Page Mail, The Free Monthly Beer News Digest for the Online Beer Enthusiast,* rbpmail@realbeer.com, July 2004, 2.
423. O'Hara, *Great American Beer, 50 Brands that Shaped the 20th Century,* 37.
424. Charles Sheehan, "It's the Bottle . . . and a Lot More," *Lewiston Morning Tribune* (Lewiston, Idaho), August 25, 2004, p. 12B.
425. Charles Sheehan, "It's the Bottle . . . and a Lot More," 12B.
426. *Real Beer Page Mail, The Free Monthly Beer News Digest for the Online Beer Enthusiast,* rbpmail@realbeer.com, July 2004.
427. *Real Beer Page Mail, The Free Monthly Beer News Digest for the Online Beer Enthusiast,* rbpmail@realbeer.com, July 2007.
428. *Real Beer Page Mail, The Free Monthly Beer News Digest for the Online Beer Enthusiast,* rbpmail@realbeer.com, September 2007.
429. *Real Beer Page Mail, The Free Monthly Beer News Digest for the Online Beer Enthusiast,* rbpmail@realbeer.com, June 2010.
430. *Real Beer Page Mail, The Free Monthly Beer News Digest for the Online Beer Enthusiast,* rbpmail@realbeer.com, August 2010.

Glossary of Brewing Terms Used in This Volume[1]

Adjuncts. Various grains, sugars, or syrups added to barley malt in the mashing stage of the brewing process. Corn and rice are the most common.

Alcohol content. The amount of alcohol contained in a beverage. In beer, the specific gravity of the wort before fermentation relates to the final amount of alcohol in the beer. Traditionally, beer was limited to 16 percent alcohol, which was rarely achieved because the high alcohol level killed the yeast. Recent research has developed yeast that can produce much higher alcohol levels.

Alcohol in a final product may be measured by volume or weight. Alcohol weighs less than water, so alcohol level by weight is about 20 percent lower than alcohol by volume. Beer that is 3.2 percent by weight is about 4 percent by volume or 8 proof.

Alcoholic beverage. Any potable beverage containing ethanol (ethyl alcohol) produced by fermentation or by distillation after fermentation has taken place.

Ale. A beer brewed with a top fermenting yeast. Types include bitter ale, brown ale, India Pale Ale (IPA), and others. Historically, in England ale was a beer without hops.

Amylase. This enzyme, or biological catalyst, is a by-product of the germination process and converts the starch in grain into sugar. Sugar can be fermented by yeast but starch cannot. Brewers use the term interchangeably with diastase.

Attemperators. Boxes that brewers filled with ice and floated in the wort to cool it and to secure a uniform fermentation throughout the tub.

Barley. An ancient cereal grass or grain *(Hordeum vulgare)* of the family *Gramineae,* which has been considered ideal for brewing from the time of the earliest Western civilizations.

Barley wine. Historically, the name for wine made from barley, or beer. In modern times the term is applied to exceptionally strong (high alcohol) beers, and is a distinct category in beer judging competitions.

Beer. A beverage made by fermenting grain extracts. Dozens of varieties exist.

Bock beer. A dark, slightly sweet beer of German origin in the lager beer tradition. It was traditionally released at Easter each spring.

Bottom fermentation. To achieve fermentation, yeast is used that sinks to the bottom of the wort to work. Mentions of this are known from hundreds of years ago but the modern practice dates from southern Germany in the 1820s.

Breweriana. Collectible artifacts and antiques that were first made for the brewing industry, particularly for advertising beer.

Bung. The hole in a keg or cask through which it was filled. A cork or stopper had to be put into the bung hole. Sometimes a bushing around the hole made it stronger and assured a tight seal.

Cooperage. A workshop that makes wooden barrels.

Copper. A synonym for a brew kettle, which developed because the kettles were made of copper.

Croze. A groove cut near the ends of barrel staves to allow the round end of the barrel to fit tightly.

Decoction. Use of hot water to extract soluble material from a plant substance.

Eighteenth Amendment. The addition to the United States Constitution that outlawed "intoxicating liquor." It was ratified in 1919 and went into effect in 1920. Alcoholic beverages had previously been outlawed in many states and counties, and on the national level as an attempt to save grain during World War I.

Fermenters. Vessels in which the yeast is allowed to work on the wort. In the past these were open, but in modern times they are generally closed to keep out airborne yeast.

Fining. The technique used in the final stage of fermentation to clarify the beer by removing solid particles of yeast or other impurities. Putting the beer in a cask with wood chips was one technique; the addition of isinglass, a gelatinous substance derived from the air bladders of fish, was another.

Floor-malting. A pre-mechanical technique in which barley was saturated and then spread on a floor and allowed to sprout. The barley had to be turned many times to assure even sprouting and prevent mold and rot. This laborious technique began to be replaced by pneumatic malting in the 1880s.

Gambrinus, King. A legendary king who invented beer. Jan Primus (Johann the First), a twelfth century Duke of Brabant, may have been the namesake of the legend.

Germination. The sprouting of the barley pearls during the malting process.

Green malt. Sprouted barley that has not yet been kiln dried.

Growler. A metal pail, or in modern times, a glass jug used to carry draft beer home from a local retailer.

Hop-Jack. A device that filters the hops out of the wort before fermentation. It may also be called a hop-back or hop strainer.

Hops. The flowers of the vine *Humulus lupus* used in brewing to offset the sweet flavor of the malt, and to add flavor, aroma, and character, and prevent spoilage in the beer.

Hydrometer. A glass instrument used to measure the specific gravity of liquids. A float rests inside a tube, which compares a given liquid to the gravity of water. Alcohol in beer is measured with such a device.

Internal Revenue Act of 1862. This federal law placed a tax of $1 per barrel on beer. The records of tax collection thus generated provide lists of legal brewers from that time through the twenty-first century.

Kilning. In the final stage of malt production, dry heat is used to dry and roast the barley. The temperature and duration of the kilning determines the final color of the malt.

Krausening, also kraeusening. The addition of wort that is still fermenting to aged beer to create a secondary fermentation that carbonates the beer before consumption.

Lager beer. A beer made with bottom fermenting yeast. First made in America about 1840, it is lighter and more effervescent than ale. The name means "storage" in German.

Lagering. Lager beer is stored in a cold room in large casks for several weeks to reach a peak of flavor. The need for lagering led to the need for the underground cellars found at many American breweries of the nineteenth century.

Lauter tun. A tub with a false, perforated bottom designed to filter the wort. The solid pieces of grain settle to the bottom in the tun, and then the wort is drained through this mass, which filters it. The wort would then go to the brew kettle for boiling.

Lautering. The wort is strained to remove the small pieces of solid, spent grain. This grain is an excellent animal feed and can be used in baking.

Local option. These laws allowed cities or counties to unilaterally outlaw liquor while the larger political unit still allowed it.

Lovibond, Degrees. Maltsters in America in the late twentieth century measured the degree of the roast on the malt—how light to dark the finished malt was—by degrees Lovibond. In Europe, the measure was the European Brewing Convention scale.

Malt. Grain (usually barley) that has been sprouted, then dried and cleaned. This results in a higher sugar content, less starch, and the presence of enzymes. It is the basis for brewing.

Malting. The process of turning barley or other grains into malt. The steps include steeping in water, germination, then kiln drying.

Maltose. The type of sugar that forms in barley from the action of amylase during germination.

Maltster. A person who performs the process of turning raw barley into malt.

Mash. A mixture of ground grains and water; this allows the grain sugars to dissolve into the liquid. In brewing, this is generally called by its German name, wort.

Mash tun (mash tub). A round vessel in which the ground grain and water are heated and stirred to dissolve the sugar out of the grain. Mechanical stirring devices and false bottoms to assist in pouring off the final product were developed to replace hand labor.

Monolith. A mixture of sawdust, mineral carbonates, and chlorides used to make floors on which to malt barley.

Near beer. A beverage made in the same manner as beer but with a very low alcohol content due to the type of yeast used or the process of de-alcoholizing or other adjustments. The name became popular after 1900.

Pasteurization. The process of heating beer or other products after they are manufactured in order to kill microorganisms. It is named for its inventor, Louis Pasteur. The popularity of bottled beer dates from the introduction of this technique for beer.

Pfaff. A tube or pipe that infuses water into the bottom of the mash tun.

Pneumatic malting. A process of forcing air through wet barley that is germinating. Temperature can thus be controlled and hand turning is minimized.

Porter. A top-fermented beer developed in England that is dark, heavy, and a little sweet.

Pure Food and Drug Act, 1906. This law raised the question of whether adjuncts such as corn, rice, or sugar were adulterations to beer and thus deserved to be illegal. These were approved but other substances were outlawed.

Racking. Transferring beer from storage vats to barrels to be sent out of the brewery. The central problem was to keep the carbonation in the beer and avoid spillage.

Ruh. The storage time for bottom fermentation beers. This can last from a few days to a few weeks in a cold, but not freezing, room. Ruh casks are the huge barrels in which the aging takes place.

Rundbogenstil. The rounded, arched style of architecture popular in U.S. breweries built from approximately 1860 to 1910, and derived from German styles.

Saccharometer. A device that measures the amount of sugar in a solution by measuring the specific gravity. These were introduced about 1789 and underwent constant improvement.

Small beer. A British name for weak beer.

Sparger. A device to spray water into a mash tub.

Sparging. After the mash tub is emptied, hot water is sprayed over the spent grain in the bottom of the mash tun to rinse out the remaining sugar.

Steam beer. A Pacific coast development in which lager was fermented at unusually high temperatures due to the lack of ice to cool the storage areas. The beer was unique and had very high pressure in the kegs.

Stout. A top-fermented beer that is very dark and strong and highly hopped.

Twenty-first Amendment. An alteration to the United States Constitution that repealed the Eighteenth Amendment and allowed states to regulate liquor. This was ratified in December, 1933. Beer had been declared non-intoxicating and was no longer outlawed at the federal level after April of that year.

Tun. A barrel used in brewing beer with approximately a 250-gallon capacity.

Underback. The tub into which the wort runs after being drained from the mash tun (lauter tun). This tub was also known as a grant.

Volstead Act. A refinement of the Eighteenth Amendment, which outlawed intoxicating liquor without defining it. This law established the one-half of 1 percent standard as the highest level permitted in a nonalcoholic product.

Wort. Pronounced *wurt*, this is the sugary liquid that results from malted grain being steeped in hot water. An old saying declares that the brewer makes the wort, but the yeast makes the beer.

Yeast. A microorganism that was not understood until the scientific age. It produces zymase, which is the primary agent in alcoholic fermentation. Essentially, yeast eats sugar and produces alcohol, carbon dioxide, and heat.

Zymology. The study of fermentation.

NOTES

1. For more information on these terms, see Dan Rabin and Carl Forget, *Dictionary of Beer and Brewing: The Most Complete Collection of Brewing Terms Written in English* (Boulder, Colorado: Brewers Publications, 1988) or William L. Downard, *Dictionary of the History of the American Brewing and Distilling Industries* (Westport, Conn.: Greenwood Press, 1980).

Organizations That Assist in Identifying Brewery Artifacts

American Breweriana Association

The American Breweriana Association is a nonprofit, tax-exempt IRS 501 (c)(3) educational and historical corporation organized to advance the public knowledge of brewing and breweries and dedicated to the preservation of the memories and artifacts of America's historic breweries. In partnership, the organization has restored the brewery at Potosi, Wisconsin, and transformed it into the National Brewery Museum. *The American Breweriana Journal* is the official publication of the American Breweriana Association.

Contact Information
P.O. Box 595767
Fort Gratiot, MI 48059–5767
Telephone: 810-385-7101
Fax: 810-385-7121
Website: www.americanbreweriana.org/

Brewery Collectibles Club of America

(formerly the Beer Can Collectors of America)
As quoted from the club's website, "This is the place to learn about the brewery collectible hobby, how it started, why we do it and how you can too. Learn a bit of beer can history while you're at it." The club publishes *Beer Cans and Brewer Collectibles Magazine* and the two-volume *United States Beer Cans*. It has annual national meetings and local chapters.

Contact Information
747 Merush Court
Fenton, MO 63026–2092
Telephone: 636-343-6486
Fax: 636-343-6436
www.bcca.com/

Eastern Coast Breweriana Association

The Eastern Coast Breweriana Association, or ECBA, is dedicated to the preservation of the history of the American Brewing Industry through breweriana collectibles. ECBA, the oldest breweriana organization, supports local and regional breweries and believes that breweriana is the key to understanding the American brewing history.

Contact Information
P. O. Box 1392
Doylestown, PA 18901
www.eastcoastbrew.com

The National Association of Breweriana Advertising

As quoted from the group's website, "The National Association of Breweriana Advertising (NABA) was organized in 1972 to encourage the collection, preservation and study of American brewery advertising on a national level. While the founding members were mostly from Wisconsin and Illinois, and our Membership is still heaviest in the Midwestern states, NABA soon developed into a national organization. NABA is incorporated in Illinois as a non-profit association, and is directed by Officers and a Board elected by the Membership. Members' collections range from pre-Prohibition items to current commemorative brewery steins and advertising and packaging items from the burgeoning microbrewery industry, and are as varied in content and scope as the diversity of the brewery industry's advertising and packaging. Collection emphasis can be from a single brewery or town, or a single category or type of advertising (or packaging) to those that proudly feature 'All Breweriana!'

The Members themselves are of both sexes, all ages, are beer lovers or non-drinkers; are 'just' collectors, members of historic brewery families, or antique dealers, show promoters, or people active in one phase or another of the home or commercial brewing industry. Many member families now include a second generation of active collectors, with several substantial collections built by youngsters that seem to favor coasters, crowns, sports schedules and the like. A strong code of ethics is encouraged by NABA, and should govern all dealings between members. There is continuous effort to alert the membership on the output of fake items, and the organization strives to curb the production and distribution of such items.

"A quarterly publication, *The Breweriana Collector*, contains articles and features on breweriana, reports and comments on the status of the hobby and the brewing industry, collecting tips and hints, a beer tasting column, Reproduction Alerts when needed, notices of shows and conventions of interest, and short ads by members. Photos printed in full color add greatly to the enjoyment and appreciation of interesting brewery collectibles, and for years *The Breweriana Collector* has followed this rule.

"An annual Membership Directory listing the entire membership, alphabetically and geographically, has become a 'Who's Who' of the breweriana collecting hobby."

Contact Information
P.O. Box 64
Chapel Hill, NC 27514–0064
www.nababrew.com/

National Brewery Museum

Located in the Potosi Brewery Complex
209 S. Main Street
Potosi, WI
Mailing address:
ABA Museum Foundation
P.O. Box 507
Platteville, WI 53818–0507
www.nationalbrewerymuseum.org/

Archaeological Excavations of Breweries and Related Sites for Comparative Purposes

Site Name: **Aztec Brewery**[1]
Location: San Diego, California
Dates: 1930s, dug 1990
Excavator: RECON Co.
Site Type: Brewery
Major Questions and Findings: The excavators found areas where machines could be mounted on brick or concrete foundations.

Site Name: **Klug Brewery**[2]
Location: Challis, Idaho
Dates: 1879–82, dug 1987
Excavator: North Custer County Historical Society
Site Type: Brewery
Major Questions and Findings: Building was stabilized and stone construction techniques surveyed (figs. AIV.1 and AIV.2).

Site Name: **Harper's Ferry**[3]
Location: West Virginia
Dates: c.1885–1890, dug 1988
Excavator: National Park Service
Site Type: Brewery
Major Questions and Findings: Physical evidence from recovered bottles was applied to questions of labor and social conditions.

Site Name: **Arell's Tavern**[4]
Location: Alexandria, Virginia
Dates: c.1700s, dug 1990s
Excavator: Gadsby's Tavern Museum
Site Type: Tavern
Major Questions and Findings: Posset pots recovered indicate early use of beer in mixed drinks in a working class setting.

Site Name: **Ice Harvest Industry**[5]
Location: Houghtaling Island, Greene County, New York
Dates: c.1900, dug 1998
Excavator: Corps of Engineers
Site Type: Ice Houses
Major Questions and Findings: A seasonal work force of 20,000 served 135 ice houses on the Hudson River.

Site Name: **Saloons of Virginia City, Nevada**[6]
Location: Virginia City, Nevada
Dates: c.1860s to 1880s, dug c.2000
Excavator: Donald L. Hadesty, Kelly J. Dixon, and Nevada State Historic Preservation Office
Site Type: Saloons
Major Questions and Findings: Beer containers and dispensing hardware were found in four saloons with different social clientele.

Site Name: **Hop houses and brewery**[7]
Location: Genesee Country Village and Americana Village, New York
Dates: unknown
Excavator: N/A
Site Type: Hop houses and brewery
Major Questions and Findings: Structures moved to these sites and those rebuilt there give physical examples of breweries and hop kilns of the early national period.

Figure AIV.1. Challis, Idaho, brewery in 1986. (Photo by author.)

Figure AIV.2. Challis, Idaho, brewery showing the monolithic window lintels. (Photo by author.)

Site Name: **Single Brothers Brewery/Distillery**
Location: Salem, North Carolina
Dates: 1773, dug 1987–88
Excavator: Industrial Complex Archaeology Sites[8]
Site Type: Brewery
Major Questions and Findings: Excavations of 1987–88 uncovered the site, which is now protected.

Site Name: **Portner's Brewery**
Location: Alexandria, Virginia[9]
Dates: 1860s, dug 1999–2000
Excavator: Parsons Engineering Science, Inc.
Site Type: Brewery
Major Questions and Findings: Excavations have unearthed a Civil War-era lager brewery.

Site Name: **Brewery Flats**
Location: Minneapolis, Minnesota[10]
Dates: 1860s to 1910, dug in 1983, 1989
Excavator: Scott F. Anfinson, Minnesota Historical Society
Site Type: Brewery
Major Questions and Findings: Sites of four early breweries were uncovered and mapped.

Site Name: **O'Brien Brewing and Malting**[11]
Location: Klondike City, Yukon, Canada
Dates: 1904 to 1919, dug 1998
Excavator: David V. Burley & Michael Will, Department of Archaeology, Simon Fraser University
Site Type: Brewery
Major Questions and Findings: The failure of a plant in the far north with high ambitions was revealed.

Site Name: **The Great Baltimore Brewery Dig**[12]
Location: City of Baltimore
Dates: 1780s to 1870s, dug 1983
Excavator: City of Baltimore, Center for Urban Archaeology
Site Type: Brewery
Major Questions and Findings: The highly publicized dig was kept visible as part of the historical interpretive park in the city.

Site Name: **Law Brewery Cave**[13]
Location: Dubuque County, Iowa
Dates: 1860s, studied 1999
Excavator: Department of Transportation and University of Iowa, Marlin R. Ingalls
Site Type: Brewery
Major Questions and Findings: The University of Iowa evaluated a cave used to age beer and as a beer garden, and later had other uses.

Site Name: **Quindaro Brewery**[14]
Location: Quindaro, Kansas
Dates: 1857
Excavator: Larry J. Schmits
Site Type: Brewery
Major Questions and Findings: The cave was compared with the Law Cave. Construction techniques were also studied.

Site Name: **Shuter's Hill Brewery**[15]
Location: West End, Alexandria, Virginia
Dates: Burned 1893, dug 1993–1994
Excavator: Parsons Engineering Science, Inc.

Figure AIV.3. Steps leading into the Schneider Beer Vault. (Photo courtesy of Hicks & Co., Austin, Texas.)

Figure AIV.4. End walls and ceiling of the Schneider beer vault. (Photo courtesy of Hicks & Co., Austin, Texas.)

Site Type: Brewery
Major Questions and Findings: This other brewery in Alexandria allowed comparisons with the more successful Portner's Brewery.

Site Name: **Schneider Beer Vault**[16]
Location: Austin, Texas
Dates: dug 1999–2001
Excavator: James W. Karbulp, Hicks & Company
Site Type: Brewery
Major Questions and Findings: The vault was built but never the ground-level brewery due to a death in the family. The vault was designed for a brewery but never used for that purpose. Construction techniques were discovered (figs. AIV.3 and AIV.4).

Site Name: **Jacob Dobmeier Brewery Complex**[17]
Location: Central Park, Grand Forks, North Dakota
Dates: Nineteenth century
Conductor of Study: Archaeo-Physics, LLC, Minneapolis, MN
Site Type: Brewery
Major Questions and Findings: Electrical resistivity studies detected buried structures associated with the brewery.

Site Name: **Gilbert Brewery and House**[18]
Location: Virginia City, Montana
Dates: 1864 to c.1920
Excavator: Montana Heritage Commission
Site Type: Brewery
Major Questions and Findings: The brewery owner's house and the extant brewery were studied and preserved (fig. AIV.5). No actual excavation.

Figure AIV.5. Sign on the Gilbert Brewery, Virginia City, Montana. (Photo by author.)

Site Name: **The Kreische Brewery**[19]
Location: Fayette County, Texas
Dates: 1882–1894, dug in 1978, 1979, 1982, 1985
Excavator: Todd McMakin, Texas Parks and Wildlife, and James E. Corbin, Stephen F. Austin State University
Site Type: Brewery
Major Questions and Findings: Structure and the implements of brewing were studied.

Site Name: **The Kun-Rudolph Brewery**[20]
Location: Springfield, Illinois
Dates: 1850s to 1870s, dug 1993
Excavator: Floyd Mansberger, Illinois Dept. of Transportation
Site Type: Brewery
Major Questions and Findings: Investigated the layout of associated structures, especially subterranean structures.

Site Name: **The City Brewery Site**[21]
Location: Des Moines, Iowa
Dates: 1855 to 1884, dug 1995–96
Excavator: Leah D. Rogers, Des Moines Metropolitan Transit Authority
Site Type: Brewery
Major Questions and Findings: Utilization of space at the brewery was studied.

NOTES

1. Judy D. Tordof, "Current Research: Pacific West: California," *Society for Historical Archaeology Newsletter* 26, no. 3 (October 1993): 30.

   The Rathskeller of the Aztec Brewery with its mural depicting a Pre-Columbian ritual can be seen in Will Anderson, *Beer, USA*, 1986, 65.

   "Archaeological First," *American Breweriana Journal* (January-February 1994): 16.
2. Molly O'Leary Cecil, "City Works to Restore Old Brewery," *The Post-Register* (Idaho Falls, Idaho), July 30, 1987, p.1, c. 1–4. "Brewery Renovation Brings Back History to Historic Challis," *The Spokesman-Review* (Spokane, Washington), October 19, 1987, A8, c. 1–2.
3. Deborah A. Hull-Walski and Frank L. Walski, "There's Trouble a-Brewin': The Brewing and Bottling Industries at Harper's Ferry, West Virginia," *Historical Archaeology* 24, no. 4 (1994): 106–121. Hull-Walski and Walski, "Brewing and Bottling in Harper's Ferry, West Virginia," Chapter 17 in Paul A. Shackel, ed., and Susan E. Winter, "Principal Investigator, Interdisciplinary Investigations of Domestic Life in Government Block B: Perspectives on Harpers Ferry's Armory and Commercial District," Occasional Report No. 6, Regional Archaeology Program, National Capital Region, National Park Service, U.S. Department of the Interior Washington, D.C., 1993.
4. Candace Beane and Ellen Donald, "Archaeology Sets The Tavern Table: Alexandria Tavern Ceramics, 1760 to 1820" (Alexandria, Virginia: Gadsby's Tavern Museum, 1994), not paginated.

5. David Starbuck, "Current Research: Northeast: Ice Harvest Industry," *The Society for Historical Archaeology Newsletter* 31, no. 3 (Fall 1998): 13.
6. Kelly J. Dixon, *Boomtown Saloons: Archaeology and History in Virginia City* (Reno: University of Nevada Press, 2005).
7. Tomlan, *Tinged with Gold: Hop Culture in the United States*, 219.
8. Kathleen Gilmore, "1979 The Single Brothers' Industrial Complex: Research Plan, Old Salem, North Carolina," The Conference on Historic Site Archaeology Papers 1978 13:76–108.
9. "Portner's Brewery," Alexandria Archaeology Museum, Preservation in Action, City of Alexandria, Virginia, Nov. 28, 2010. http://oha.alexandriava.gov/archaeology/pia-sites/pia_portners_brewery.html.
10. Scott F. Anfinson, "Archaeological Site Inventory: Brewery Flats," and "Archaeology of the Central Minneapolis Riverfront I," *The Minnesota Archaeologist* 48, nos. 1 and 2 (1989).
11. Burley and Will, "Special Brew: Industrial Archaeology and History of the Klondike Brewery." Burley and Will, " 'The Beer That Made Klondike Famous and Milwaukee Jealous': The O'Brien Brewing and Malting Company Site, Klondike City, Yukon," 37–54.
12. Elizabeth Anderson Comer, "Public Archaeology as a Tool for Community Preservation and Empowerment," *Archaeology* 28 (Nov. 2010), http://eacarchaeology.com/publicBalto.html.

    Douglas Sanford, "Current Research: Mid-Atlantic: Baltimore," *The Society for Historical Archaeology Newsletter* 26, no. 3 (October 1993): 15.
13. Marlin R. Ingalls, "A Phase II Historic Architectural Evaluation of The Law Brewery Cave, Site 31-00008 Primary Roads Project STP-32-1 (1)—2C-32 a.k.a. PIN 95-31080-1 Dubuque County, Iowa." Office of Project Planning Iowa Department of Transportation, Ames, Iowa 50010, Carl A. Merry and Stephen C. Lensink Co-Principal Investigators, Project Completion Report Volume 22, Number 16, 1999, p. 19.
14. Larry J. Schmits, "Quindaro: Territorial Free-State Port on the Missouri River," *Missouri Archaeologist* 49 (1988): 121.
15. Mark Walker, Timothy Dennee, and Brian Crane, "The Receptacles Were Emptied of Their Contents:" Archaeological Testing of Area II-B of the Carlyle Property and Excavation of the Shuter's Hill Brewery Site (44AX35), Alexandria Virginia." Parsons Engineering Science, Inc. Fairfax, Virginia, 1996. Shirley Scalley, "Shuter's Hill Brewery," Alexandria Archaeology Museum, Preservation in Action, no. 35 (Nov. 28, 2010), < http://oha.alexandriava.gov/archaeology/pia-sites/pia_shuters_hill.html> .
16. Rachel Feit, James W. Karbula, John Clark, and S. Christopher Caran, with contributions by J. Andrew Moreman, Timothy Griffith, and Jonathan Jarvis, "Boarding Houses, Bar Rooms and Brothels—Life in a Vice-District: Archaeological Investigations of A Changing Urban Neighborhood in Austin, Texas." Volume I: Historical Background. Volume II: The Results of Investigations. Hicks & Company, Archeology Series No. 104, 2003. James W. Karbula, Principal Investigator.
17. "Shallow Subsurface Geophysical Survey, Central Park 19th Century brewery complex, Grand Forks, North Dakota," Archaeo-Physics internet, Nov. 28, 2010, http://www.archaeophysics.com/central.
18. Perry Backus, "Virginia City Striving to Preserve What's Left of Brewery," *The Montana Standard* (Butte, Montana), January 4, 2004.
19. Todd McMakin and James E. Corbin, compilers, "The Kreische Brewery State Historic Site (41FY128): An Archeological and Architectural Assessment." Report prepared for Texas Parks and Wildlife, Texas Antiquities Permit 173, October 2001.
20. Mansberger, "The Industrial Archaeology of Breweries: Archaeological Investigations at the Mid-Nineteenth Century Kun-Rudolph Brewery, Springfield," 1.
21. Rogers, " 'It was Some Brewery': Data Recovery of the City Brewery Site, 13PK661, Des Moines, Iowa."

BIBLIOGRAPHY

Anderson, Will. *Beer, USA*. Dobbs Ferry, New York: Morgan & Morgan, 1986.

This nontechnical book is essentially a time line of brewing history with unusual information and excellent illustrations.

Appel, Susan K. "New Finds in Brewery Architecture." *The Breweriana Collector* 7 (Spring 1999).

Describes the business archives of a brewery architect that are publicly available in Chicago.

Appel, Susan K. "Chicago and the Rise of Brewery Architecture." *Chicago History* 24 no. 1 (Spring, 1995): 4–19.

Chicago was a major brewing center at one time and had more brewery architects than any other city.

Appel, Susan Kay Bigley. "The Midwestern Brewery Before Prohibition: Development of an American Industrial Building Type," dissertation, University of Illinois at Urbana-Champaign, 1990.

Discusses the development of the architecture in St. Louis and Cincinnati in particular. No other source has so much brewery architecture explained and analyzed.

Appel, Susan K. "Artificial refrigeration and the Architecture of 19th-Century American Breweries." *The Journal of the Society for Industrial Archaeology* 16, no. 1 (1990): 21–38.

Refrigeration was possibly the most important technical advance in brewing in the nineteenth century.

Apps, Jerry. *Breweries of Wisconsin*. Madison: University of Wisconsin Press, 1992.

Wisconsin was a major brewing state with larger national shippers and many local brewers. It also has many breweries that remain standing.

"Archaeological First." *American Breweriana Journal* (January–February 1994): 16.

Anonymous blurb describes archaeological work on the Aztec Brewery of San Diego.

Arnold, John P., and Frank Penman. *History of the Brewing Industry and Brewing Science in America, Prepared as Part of a Memorial to the Pioneers of American Brewing Science: Dr. John E. Siebel and Anton Schwarz*. Chicago: G. L. Peterson, 1933.

This is a thorough history of the brewing industry by men who worked in the profession. The book's publication date corresponds with the time when brewers were trying to upgrade their status as Prohibition was being reconsidered.

Babel, Dave. "Beer Can Closeup." *Beer Cans and Brewery Collectibles* (December 2006/January 2007): 36–37.

Discusses dating of special cans and includes unusual information on them.

Backus, Perry. "Virginia City Striving to Preserve What's Left of Brewery." *The Montana Standard*, January 4, 2004.

Newspaper article about the work being done on the old brewery of this Montana town.

Baron, Stanley Wade. *Brewed in America: A History of Beer and Ale in the United States*. New York: Arno Press, 1972 (copyright 1962).

This was the standard history of brewing in America and gives a broad, solidly researched view up to that year.

Barth, Heinrich Joh., Christiane Klinke, and Claus Schmidt. *The Hop Atlas: The History and Geography of the Cultivated Plant*. Nuremberg, Germany: Joh. Barth & Sohn, 1994.

Details hop growing areas all over the world including the United States. Hops have always been sold on international markets.

Beane, Candace, and Ellen Donald. "Archaeology Sets The Tavern Table: Alexandria Tavern Ceramics, 1760 to 1820." Alexandria, Virginia: Gadsby's Tavern Museum, 1994.

This pamphlet describes the ceramics collection from more than 175 sites in the city and shows tableware that was used where beer was retailed.

Blum, Peter H. *Brewed in Detroit: Breweries and Beers Since 1830*. Detroit: Wayne State University Press, 1999.

A brewing executive offers the kind of insight only one with his background could offer. This book is especially good with regard to the 1990s era of consolidation which affected Stroh, his employer.

"Bottle Language." Washington, D.C.: Glass Packaging Institute, n.d.

Pamphlet shows and explains maker's marks, dates, and locations embossed on glass containers.

Brande, W. *The Town and Country Brewery Book; or Every Man his Own Brewer*. London: Dean and Munday, c.1830. Reprinted, Raudins Publishing, 2003.

A practical guide for brewing as practiced in 1830. Especially valuable for information on the bizarre assortment of additives which were apparently common in Britain at the time.

Brown, Sanborn C. *Wines and Beers of Old New England: A How-To-Do-It History*. Hanover, New Hampshire: University Press of New England, 1978.

A rich source of information on brewing and fermenting of all types during Colonial and early National days. It addresses home brewing, not commercial enterprise.

Bruman, Henry J. *Alcohol in Ancient Mexico*. Salt Lake City: University of Utah Press, 2000.

This version of Bruman's dissertation covers pre-Columbian fermented beverages made in each small region of Mexico.

Bull, Donald, Manfred Friedrich, and Robert Gottschalk. *American Breweries*. Trumbull, CT: Bullworks, 1984.

This is the second reference book to list all known breweries in America for its entire history. A number of breweries managed to avoid the historical record, but this is the place to start to find any former brewery.

Burden, Paul. *The Guide to U.S. Painted Label Beer Bottles*. Dover, MA: Norfield Publishing, 1990.

In addition to embossing and paper labels, bottle labels have been and still are painted on.

Burley, David V. and Michael Will. "Special Brew Industrial Archaeology and History of the Klondike Brewery." *Occasional Papers in Archaeology* 11. Heritage Branch Government of the Yukon, *Hude Hudan* Series, 2002.

Describes the archaeology done at the site.

Burley, David V., and Michael H. Will. " 'The Beer That Made Klondike Famous and Milwaukee Jealous': The O'Brien Brewing and Malting Company Site, Klondike City, Yukon." *Industrial Archaeology* 26 no. 1 (2000): 37–54.

A description of this far-north brewery.

Busch, Jane. "An Introduction to the Tin Can." *Historical Archaeology* 15 no. 1 (1981): 95–104.

Describes the place of the beer can in the whole history of canned food.

Byrn, M. L. *The Complete Practical Brewer*. Philadelphia: Henry Carey Baird, 1852. Reprint, Chagrin Falls, OH: Raudins Publishing, 2002.

Byrn was an M.D. who claimed to have lived on the Mississippi River for several years. His explanation of beer and breweries precedes the surge in popularity of lager beer.

Cecil, Molly O'Leary. "City Works to Restore Old Brewery." *The Post-Register*, Idaho Falls, Idaho, July 30, 1987, p.1 , c. 1–4.

Describes the archaeology of the Challis, Idaho, brewery.

Challis Brewery Historic District, United States Department of the Interior National Register of Historic Places Inventory —Nomination Form, 1979.

This stone building was a brewery for a few years, then experienced a variety of uses. In the early 1980s, it was a historical museum. Archaeological work has been done on the premises, and it is still extant.

Cochran, Thomas C. *The Pabst Brewing Company: The History of An American Business*. New York: New York University Press, 1948.

This is the only scholarly history of a single American brewery from its inception to the date of the writing. Cochran seems almost apologetic for selecting a brewery for study.

Colyer, Frederick. *Breweries and Maltings: Their Arrangement, Construction, Machinery and Plant*. 2nd ed. London: E. & F. N. Spon, 1880.

Although this is a British book, Colyer covered every practical aspect of building a state-of-the-art 1880 brewery. He also listed British patents relating to malt and hops from 1635 to 1879, and included drawings of breweries from every angle.

Combined History of Randolph, Monroe and Perry Counties, Illinois. Philadelphia: J. L. McDonough & Co., 1883.

Describes Kaskaskia, site of an early brewery on the Mississippi River.

Cooler, Kathleen E. Hudson. "Hop Agriculture in Oregon: The First Century," thesis, Portland State University, 1986.

The history of hop growing in an area where it was of major importance.

Creighton, Gregg B. "Schade Brewery Restoration: Preservation of a Prototype," thesis, University of Idaho, 1984.

Plan for reuse of a Spokane, Washington, brewery.

Cunliffe, Barry. "Afterward: Historical Archaeology in the Wider Discipline," in *The Cambridge Companion to Historical Archaeology,* eds. Dan Hicks and Mary C. Beaudry. Cambridge: Cambridge University Press, 2006, 314–319.

Day, Dan. "Brewery Becoming Co-op Apartments," *The Spokesman -Review,* Spokane, Washington, February 13, 1983, p. E2.

The development of the old Blatz brewery complex in Milwaukee.

Dick, Charles Edwin. "A Geographical Analysis of the Development of the Brewing Industry of Minnesota," dissertation, University of Minnesota, 1981.

Explores how population and ethnicity pertain to brewery locations.

Dixon, Kelly J. *Boomtown Saloons: Archaeology and History in Virginia City*. Reno: University of Nevada Press, 2005.

Describes excavations at several saloons in this Nevada town.

Downard, William L. *Dictionary of the History of the American Brewing and Distilling Industries*. Westport, Conn.: Greenwood Press, 1980.

This is an excellent source to understand the specialized vocabulary of brewing.

Downard, William L. *The Cincinnati Brewing Industry, A Social and Economic History*. Oberlin: Ohio University Press, 1973.

Very professional and well executed volume by a professor of history.

Driver, Harold E. "Alcoholic Beverages in Native North America." *Indiana Academy of Science* 64 (1955): 50–51.

A good synopsis of the brewed beverages of this region.

Ehret, George. *Twenty-Five Years of Brewing with an Illustrated History of American Beer Dedicated to the Friends of George Ehret*. New York: George Ehret, 1891.

Ehret was a New York City brewer and one of the most successful in America. He explains the business and the physical aspects of his brewery with great pride.

Flanagan, John T. *Theodore Hamm in Minnesota: His Family and Brewery*. Minneapolis: Pogo Press, 1989.

The author is the grandson of the subject, and writes with affection and scholarly standards.

Fontana, Bernard L., and J. Cameron Greenleaf, et al. "Johnny Ward's Ranch: A Study in Historical Archaeology." *The Kiva* 28 (1962): 1–2.

Bottle find at this Arizona ranch demonstrates the ubiquity of beer bottles.

Friedrich, Manfred, and Donald Bull. *The Register of United States Breweries, 1876–1976*. Two volumes. Trumbull, CT: Donald Bull, 1976.

This was the first attempt to create a list of all U. S. breweries. It was based mostly on government records, and has been updated twice by other authors.

Gaynor, Bob. "The Breweries of Burlington, New Jersey." *American Breweriana Journal* 137 (September–October 2005): 33–35.

An exploration of very early brewery sites in this city.

Gibson, Charles. *The Aztecs Under Spanish Rule: A History of the Indians of the Valley of Mexico, 1519–1810*. Stanford: Stanford University Press, 1964.

The author refers to European brewing starting in Mexico, which was the first instance of European brewing coming to the New World.

Guetig, Peter R., and Conrad D. Selle. *Louisville Breweries: A History of the Brewing Industry in Louisville, Kentucky, New Albany and Jeffersonville, Indiana*. Louisville: Mark Skaggs Press, 1995.

Good, broad-based history.

Guse-Noritake, Judy R. "The Schade Brewery, A Restoration," thesis, University of Idaho, 1978.

Plan for reuse of a Spokane, Washington brewery.

Hardman, Michael. *Beer Naturally*. London: Bergstrom and Boyle, 1976.

Part of the "real ale" movement to preserve British brewing tradition and beer styles.

Harris, Moira F. "From Limestone to Rubble at a Brewery Full of Memories." Golden Age of Brewery Architecture, *American Breweriana Journal* 118 (July–August 2002): 14–20.

Describes the destruction of the Hamm's Brewery in St. Paul.

Hennech, Mike, and Doug Pierson. *Beer Trucks: A Photographic Salute, 1870–1992*. Irving, TX: Ale Publishing Company, 1992.

This book follows beer delivery trucks as they developed over the course of the twentieth century.

Hicks, Dan, and Mary C. Beaudry, editors. *The Cambridge Companion to Historical Archaeology*. Cambridge: Cambridge University Press, 2006.

An anthology of the school of thought in the discipline at that point in time.

Hinrichs, Hans, ed. *Rahr*. Manitowoc, Wisconsin: Thomas Delehanty Inc., 1952.

A company history published by the malting company. Valuable to see the technology of malting of that era.

Historical Atlas of the United States. Centennial Edition, National Geographic, 1988.

Contains photos and drawings of hops industrial structures.

Holian, Timothy J. *Over the Barrel: The Brewing History and Beer Culture of Cincinnati, 1800 to the Present, Volume One (1800–1919)*. St. Joseph, MO: Sudhaus Press, 2000.

Study of an important brewing center.

Hull-Walski, Deborah, and Frank L. Walski. "There's Trouble a-Brewin': The Brewing and Bottling Industries at Harper's Ferry, West Virginia." *Historical Archaeology* 28, no. 4 (1994): 106–121.

Describes the history of the brewery in relation to the bottles found at other locations in the town.

Hull-Walski, Deborah, and Frank L. Walski, "Brewing and Bottling in Harper's Ferry, West Virginia." Chapter 17 in Paul A. Shackel, ed., and Susan E. Winter, Principal Investigator, *Interdisciplinary Investigations of Domestic Life in Government Block B: Perspectives on Harpers Ferry's Armory and Commercial District*. Occasional Report No. 6, Regional Archaeology Program, National Capital Region, National Park Service, U.S. Department of the Interior Washington, D.C., 1993.

Describes the archaeology at Harper's Ferry.

Idaho Tri-Weekly Statesman, Boise, Idaho, September 17, 1885, p. 1, c. 3.

News report of an attack on Chinese hop pickers by Native Americans; the Chinese had replaced the Native Americans during that season's labor.

Ingalls, Marlin R. "A Phase II Historic Architectural Evaluation of The Law Brewery Cave, Site 31-00008 Primary Roads Project STP-32-1 (1)—2C-32 a.k.a. PIN 95-31080-1 Dubuque County, Iowa." Office of Project Planning Iowa Department of Transportation, Ames, Iowa 50010, Carl A. Merry and Stephen C. Lensink Co-Principal Investigators, Project Completion Report Volume 22, Number 16, 1999, p. 19.

Describes the examination of the cave.

Janowitz, Meta F. "Indian Corn and Dutch Pots: Seventeenth-Century Foodways in New Amsterdam/New York." *Historical Archaeology* (Health, Sanitation and Foodways in Historical Archaeology, Joan H. Gesimar and Meta F. Janowitz, editors) 27, no. 2 (1993): 6–24.

Examines the place of malt and beer in the overall production, preparation, and consumption of food and drink in that time and place.

Jones, Olive R. "Commercial Foods, 1740–1820." *Historical Archaeology* 27, no. 2 (1993): 25–41.

Describes packaging and importing of many food products including alcoholic beverages.

Joyce, Rosemary. "Writing Historical Archaeology," in *The Cambridge Companion to Historical Archaeology*, edited by Dan Hicks and Mary C. Beaudry. Cambridge: Cambridge University Press, 2006, 48–68.

Karbula, James W., Principal Investigator, with Rachel Feit, John Clark, and S. Christopher Caran. "Boarding Houses, Bar Rooms and Brothels—Life in a Vice-District: Archaeological Investigations of A Changing Urban Neighborhood in Austin, Texas." *Volume I: Historical Background. Volume II: The Results of Investigations.* Hicks & Company, Archaeology Series No. 104, 2003.

Describes the brewery construction in Austin, Texas.

Kay, Bob. "Evolution of the Budweiser Label." Pamphlet published by the author, Batavia, Illinois, January, 1993.

Bottle labels can be dated to within a few years with guides such as this.

Kaye, Edward R., and Donald A. Bull. *The Handbook of Beer Advertising Openers and Corkscrews (An Alphabetical Composite)*. Sanibel Island, Florida: Edward R. Kaye, 1984.

This book offers a typology for identifying opening devices that might be found at a brewery or retail beer establishment.

Kerr, K. Austin. "The American Brewing Industry, 1865–1920." In *The Dynamics of the International Brewing Industry Since 1800*, R. G. Wilson and T. R. Gourvish, eds. New York and London: Routledge, 1998, 176–192.

"Kessler Brewery Established 1865 and How Beer is Brewed." Helena, Montana, n.d., c.1952.

This pamphlet describes the development of this company in its isolated location.

Kieninger, Helmut. "The Influence of Raw Materials and Yeast on the Various Beer Types." *Brewers Digest* 58, no. 10 (Oct. 1983): 44.

Kious, Kevin, and Donald Roussin. "The Lost Breweries of Collinsville, Illinois [part I]." *The Breweriana Collector* 128 (Winter 2004/2005): 16–25.

Kious, Kevin, and Donald Roussin. "The Lost Breweries of Collinsville, Illinois [part II]." *The Breweriana Collector* 131 (Fall 2005): 22–25.

Kious, Kevin, and Donald Roussin. "The Lost Breweries of Collinsville, Illinois [part III]." *The Breweriana Collector* 132 (Winter 2006): 9–12.

Kious, Kevin, and Donald Roussin. "The Lost Breweries of Collinsville, Illinois [part IV]." *The Breweriana Collector* 133 (Spring 2006): 28–30.

The diary of the brewer at this small location is one of the few such documents known. A private recounting like this reveals events that the public record will never cover.

Kious, Kevin, and Donald Roussin. "William J. Lemp Brewing Company." *American Breweriana Journal* 97 (March–April 1999): 5–13.

This is a well-illustrated account of one of the most important breweries in American history.

"Know Your Malting Barley Varieties." *Brewers Digest* (May 1985): 24–25.

Describes the types of barley then grown in America and where each is popular.

Kroll, Wayne L. *Badger Breweries Past and Present*. Jefferson, WI: Wayne L. Kroll, 1976.

One of the early state brewing histories before they became popular subjects for research.

"Latter-Day Shakers." *Archaeology* (January/February 1999): 28–29.

Describes the unexpected presence of alcohol bottles in a nondrinking religious community.

Lawrence, Susan, and Nick Shepherd. "Historical Archaeology and Colonialism," in *The Cambridge Companion to Historical Archaeology*, Dan Hicks and Mary C. Beaudry, eds. Cambridge: Cambridge University Press, 2006, 69–86.

Lemmens, Ing. Gerard W. Ch. "The Breeding and Parentage of Hop Varieties." *Brewers Digest* 73, no. 5 (May 1998): 16–26.

A scholarly account by a man who works in the hop-growing profession.

Lilek, Kevin C. *United States Beer Cans with Opening Instructions, The Standard Reference Guide of Opening Instruction Flat Tops*. Aurora, Illinois: Lilek Publishing, 1998.

Photographs of more than 90 percent of all such cans.

Lilek, Kevin C. *United States Beer Cans with Opening Instructions, The Standard Reference Guide of Opening Instruction Flat Tops (Supplement)*. Aurora, Illinois: Lilek Publishing, 2006.

Includes most of the cans left out of the initial volume.

"Logistics Firm Offers Kegs for Lease." *American Breweriana Journal* 118 (July–August 2002).

Brief news note on a new business system designed to help brewers eliminate the need to locate their missing kegs.

Love, Ed, and J. Peterson Powell. *Brewery and Beer Trademarks, vol. I*. Colorado Springs: Villa Publishing, 1989.

A listing of trademarks used by breweries.

Love, Ed, and J. Peterson Powell, *Brewery and Beer Trademarks, vol. II*. Colorado Springs: Villa Publishing, 1991.

A listing of trademarks used by breweries.

Mansberger, Floyd. "The Industrial Archaeology of Breweries: Archaeological Investigations at the Mid-Nineteenth Century Kun-Rudolph Brewery, Springfield." Fever River Research, Springfield, Illinois (for Illinois Department of Transportation), 1993.

Martells, Jack. *The Beer Can Collectors Bible*. New York: Ballentine Books, 1976.

The book shows over 3,400 beer cans in color. Pull-tab variations are shown and dated.

Martells, Jack. *The Cone Top Collectors Bible*. Matteson, Illinois: Great Lakes Living Press, 1976.

The cone top is generally the most collectible can; Martells explains its history in detail.

Martin, Byron, and Vicky Martin. *Here's to Beers: Blob Top Beer Bottles, 1880–1910*. Northridge, CA: The Achin Back Saloon, 1973.

Illustrates and describes old beer bottles for collectors.

Maxwell, D. B. S. "Beer Cans: A Guide for the Archaeologist." *Historical Archaeology* 27, no. 1 (1993): 95–113.

Since 1936, most breweries have canned beer. Dating and identifying the origin of beer cans is necessary for historical archaeologists.

Maxwell, H. James, and Bob Sullivan, Jr. *Hometown Beer: A History of Kansas City's Breweries*. Kansas City: Omega Innovative Marketing, 1999.

Studies an important but less heralded brewing center.

McMakin, Todd, and James E. Corbin, compilers. "The Kreische Brewery State Historic Site (41FY128): An Archeological and Architectural Assessment." Report prepared for Texas Parks and Wildlife, Texas Antiquities Permit 173, October 2001.

A very complete examination of the entire excavation.

Meeker, Ezra. *Hop Culture in the United States, Being a Practical Treatise on Hop Growing in Washington Territory.* Puyallup, Washington Territory, 1883.

Millard, Arnold A. *History of the Carson Brewing Company: "Nevada's Oldest Business," 1860–1948*. Carson City: The Brewery Press, 1980.

Millard owned the business and described it years after. His book covers architecture and equipment from an owner's, not a technician's, viewpoint.

Miller, Carl H. *Breweries of Cleveland*. Cleveland: Schnitzelbank Press, 1998.

A well written, researched, and illustrated account of the brewing history of that city.

The Minnesota Archaeologist 48, no. 1–2. The Minnesota Archaeological Society, 1989.

This article discusses field work examining brewery building remains in the Minneapolis area.

Mittelman, Amy. "The Politics of Alcohol Production: The Liquor Industry and the Federal Government, 1862–1900," dissertation, Columbia University, 1986.

Emphasizes national policy vis-a-vis the brewing and related industries.

Morris, Danny, and Jeff Johnson. *Richmond Beers: A Directory of the Breweries and Bottlers of Richmond, Virginia*, 2nd edition. n.p.: n.p., 2000.

Charles Zoller's 1891 *Brewers' Catalogue* that is appended is very valuable for seeing the equipment for brewing of that era. The section that depicts beer bottle closures is well illustrated and shows the major types. It is a beautifully printed book.

Mundt, Diane. "Professor Planning Museum of Minnesota Brewing." *Minneapolis Star and Tribune*, August 6, 1986, p. 4B.

One of many proposed uses for abandoned breweries.

Nelson, Herbert B. "The Vanishing Hop-Driers of the Willamette Valley." *Oregon Historical Quarterly* 64, no, 3 (1963): 267–271.

Describes the old hop structures.

Newell, Diane. "With Respect to Breweries." *Historic Preservation* 27 (January 1976): 24.

A philosophical discussion of the historical significance of a much-maligned industry.

North American Wine and Spirit Journal 21, no 6 (April, 1909).

This journal for the liquor trade proudly featured new equipment and structures at several large breweries during the first decade of the twentieth century.

Nyberg, Bartell. "For Coors, the Barley Must be the Best." *Lewiston Tribune*, August 14, 1983, 2A.

Feature article about the Coors brewery.

Ogle, Maureen. *Ambitious Brew: The Story of American Beer*. Orlando: Harcourt, Inc., 2006.

This is the latest comprehensive study of the American brewing industry by an admitted nonspecialist. I found it insightful and well researched.

O'Hara, Christopher B. *Great American Beer: 50 Brands that Shaped the 20th Century*. New York: Clarkson Potter, 2006.

Gives proper credit to well-known beer brands.

O'Keeffe, Tadhg, and Rebecca Yamin. "Urban Historical Archaeology," in *The Cambridge Companion to Historical Archaeology*, Dan Hicks and Mary C. Beaudry, eds. Cambridge: Cambridge University Press, 2006, 87–103.

The authors discuss the locations of breweries in urban settings, as well as the relative preferences for beer or wine among ethnic populations of the nineteenth century.

One Hundred Years of Brewing, A Complete History of the Progress Made in the Art, Science and Industry of Brewing in the World, Particularly During the Nineteenth Century. Historical Sketches and Views of Ancient and Modern Breweries. Lives and Portraits of Brewers of the Past and Present. Chicago: H.S. Rich & Co., 1903. Reprinted by Arno Press, 1974.

This voluminous, highly illustrated book has near encyclopedic coverage of brewing history and the state of the art in 1903. It would be a basic tool for identifying breweries, brewers, and equipment, both contemporary to its publication date and since the beginning of North American brewing.

Owens, Bill. *How To Build a Small Brewery: Draft Beer in Ten Days*, 3rd ed. Ann Arbor, MI: G. W. Kent, Inc., 1992.

A good volume for succinctly explaining the basics needed to produce beer.

Polak, Michael. *Bottles: Identification and Price Guide*, 2nd ed. New York: Avon Books, 1997. Also 4th edition; Iola, Wisconsin: Krause Publications, 2002.

Beer bottles as considered by collectors.

Plavchan, Ronald Jan. "A History of Anheuser-Busch, 1852–1933," dissertation, St. Louis University, 1969.

One of the few studies of a single major brewery done by a trained scholar.

The Post Register (Idaho Falls), August 29, 1935, p. 7, c. 4.

Reports on a train load of Czechoslovakian malt brought to Idaho Falls for the first post-Prohibition beer brewing.

Powell, Stephen R. *Rushing the Growler: A History of Brewing in Buffalo*. Buffalo, NY: Apogee Design, 1996.

A history of brewing in a city not particularly noted for beer.

Rabin, Dan, and Carl Forget. *Dictionary of Beer and Brewing: The Most Complete Collection of Brewing Terms Written in English*. Boulder, Colorado: Brewers Publications, 1988.

Valuable for anyone trying to understand the specialized vocabulary of brewing.

Rejmaniak, Tom. "What Ever Happened to the Pearl Brewery Stables?" *American Breweriana Journal* 144 (November–December 2006): 15.

Renfrow, Cindy. *A Sip Through Time: A Collection of Old Brewing Recipes*. Published by author, 1995.

Beer recipes of the past, particularly early English ones, are given.

Rogers, Leah D. " 'It was Some Brewery': Data Recovery of the City Brewery Site, 13PK661, Des Moines, Iowa." Des Moines, Iowa: Des Moines Metropolitan Transit Authority, 1996.

A very detailed and well-researched account of an extensive project.

Ronnenberg, Herman W. *The Politics of Assimilation*. New York: Carlton Press, 1975.

This study traces the anti-immigrant bias frequently found in temperance speeches and literature.

Ronnenberg, Herman W. "The Sunset Brewery of Wallace, Idaho." *The Breweriana Collector* 38 (Summer, 1982): 19–22.

Gives the history of a small brewery that suffered from a forest fire, statewide prohibition, and weak management until it failed.

Ronnenberg, Herman W. "Researching Your Hometown Brewery: The Fire Insurance Connection." *The Breweriana Collector* 69 (Spring, 1990): 21–24.

This article explains the origin, location, and historical use of these maps. A great deal of physical information about breweries is available on these maps that exists nowhere else.

Ronnenberg, Herman W. "Idaho on the Rocks: The Ice Business in the Gem State." *Idaho Yesterdays*, Spring, 1990.

Shows the direct correlation between breweries and the ice cutting and retail ice business in a western state.

Ronnenberg, Herman Wiley. "Beer for the Mining Frontier, Two Breweries Served Idaho's Famed Coeur d'Alenes." *American Breweriana Journal* 45 (July–August 1990): 8–11.

A study of the brewing history in a late era mining camp that still exists in the early twenty-first century.

Ronnenberg, Herman. *Beer and Brewing in the Inland Northwest, 1850 to 1950*. Moscow, Idaho: University of Idaho Press, 1993.

The history of brewing in Idaho is covered in its many facets.

Ronnenberg, Herman Wiley. "The Palouse Washington Brewery." *The Breweriana Collector* (Spring 2003): 21–22.

Study of a small local brewery.

Ronnenberg, Herman Wiley. "The Brewery of Palouse, Washington." *Bunchgrass Historian* 32, no. 1 (2006): 4–11.

Study of a small local brewery that is partially extant.

Rorabaugh, W. J. *The Alcoholic Republic: An American Tradition*. Oxford: Oxford University Press, 1979.

Considers all alcohol consumption in America, and beer was part of it. Attempts are made to quantify alcohol usage.

Ross, William D. *An Index to American Brewers and Breweries in One Hundred Years of Brewing*. Knoxville: Campbell, Inc., 1975.

This work makes the information in that huge book much more accessible since the book itself contains no index.

Rother, Hubert, and Charlotte Rother. *Lost Caves of St. Louis: A History of the City's Forgotten Caves*. St. Louis: Virginia Publishing Co., 1996.

This study describes a couple's attempt to find and explore forgotten caves in this city. Many of these have a background as brewery caves.

Salem, F. W. *Beer, Its History and Its Economic Value as a National Beverage*. Hartford, Conn.: F. W. Salem & Company, 1880. Reprint, New York: Arno Press, 1972.

The book was written to praise the brewery industry but has solid information.

Sanford, Douglas. "Current Research: Mid-Atlantic: Baltimore." *The Society for Historical Archaeology Newsletter* 26, no. 3 (October 1993): 15.

Baltimore brewery archaeology is discussed.

Schaefer, F & M Brewing Co. *To Commemorate Our 100th Year, 1842–1942*. n.p: F & M Schaefer, 1942.

Details the state of brewing during this era.

Schluter, Hermann. *The Brewery Industry and the Brewery Worker's Movement in America*. New York: Burt Franklin, 1910. Reprinted 1970.

This remains the only attempt at a comprehensive history of the labor movement in the American brewing industry.

Schmits, Larry J. "Quindaro: Territorial Free-State Port on the Missouri River." *Missouri Archaeologist* 49 (1988): 89–145.

Details a brewery near Kansas City.

Sheehan, Charles. "It's the Bottle . . . and a Lot More." *Lewiston Morning Tribune*, August 25, 2004, p. 12B.

Aluminum bottles are introduced.

Skilnik, Bob. *The History of Beer and Brewing in Chicago, 1833–1978*. St. Paul: Pogo Press, 1999.

Covers the rich brewing history of America's second largest city.

Slade, David F. "History of the Billings Brewing Company," thesis, University of Montana, 1971.

Study of a small Montana brewery.

Slaybaugh, Kristine M. "Schade Brewery Development," thesis, University of Idaho, 1987.

Another plan for restoration of the Spokane brewery building.

Smith, Gregg. *Beer: A History of Suds and Civilization From Mesopotamia to Microbreweries*. New York: Avon Books, 1995.

A good general history of brewing.

Spokane Breweries Records, Washington State University, Pullman, Washington.

Many stacks of ledgers, letter books, and miscellaneous records from several breweries that eventually merged.

Stack, Martin Heidegger. "Liquid Bread: An Examination of the American Brewing Industry, 1865–1940," dissertation, University of Notre Dame, 1998.

Stack takes exception to the majority of standard facts and interpretations of American brewing history.

Staff of the Washington State Office of Archaeology and Historic Preservation. *Built in Washington, 12,000 Years of Pacific Northwest Archaeological Sites and Historic Buildings*. Tacoma, Washington: Washington State Historical Society, 1989.

Describes hop structures remaining in the nation's largest hop producing state.

Starbuck, David. "Current Research: Northeast: Ice Harvest Industry," *The Society for Historical Archaeology Newsletter* 31, no. 3 (Fall 1998): 13.

A brief description of the field research being conducted on sites where natural ice was harvested.

Taylor, Edith C. (Weisgerber), Philip O. Weisgerber, and Marie E. White (Weisgerber). *A Man, His Family, and His City*. n.p.: n.p., copyright Edith C. Taylor, 1982.

A local, family-oriented, study of the first brewers in Lewiston, Idaho.

Thomas, Jerry. *The Bar-Tenders Guide, or How To Mix All Kinds of Plain and Fancy Drinks*. New York: Dick and Fitzgerald, 1862.

Describes beer handling as it was practiced in saloons in the 1860s.

Tiller, Priscilla. "Hops: The Curious Crop of Lewis County." *Landmarks, Magazine of Northwest History and Preservation* 1, no. 2 (Winter 1982): 22–24.

A history of hop growing in a major hop-producing state.

Tomlan, Michael A. *Tinged with Gold: Hop Culture in the United States*. Athens: University of Georgia Press, 1992.

Describes structures and equipment used during various epochs in America.

Tordof, Judy D. "Current Research: Pacific West: Nevada." *Society for Historical Archaeology Newsletter* 31, no. 3 (Fall 1998): 25.

A report on the study of Chauncy Hashall's ice ponds in Old Washoe city, Nevada.

Tordof, Judy D. "Current Research: Pacific West: California." *Society for Historical Archaeology Newsletter* 26, no. 3 (October 1993): 30.

A brief report on the archaeological study of the Savage Tire Factory and Aztec Brewery excavation in San Diego, California.

Toulouse, Julian Harrison. *Bottle Makers and Their Marks*. New York: Thomas Nelson Inc., 1971.

Helps identify the manufacturers of numerous bottles.

Van Wieren, Dale P. *American Breweries II*. West Point, PA: East Coast Breweriana Association, 1995.

The latest revised listing of known breweries in America throughout history.

Vance, Del. *Beer in the Beehive: A History of Brewing in Utah*. Great Salt Lake: Dream Garden Press, 2006.

History of brewing in that state written by a commercial brewer.

Vetter, John. "What is Unique About These Crowns?" *American Breweriana Journal* 144 (November–December 2006): 36, 42.

Explains the technological change that occurred when the Painter closure adapted to "short skirt" design and analyzes the savings in metal that resulted.

Walker, Mark, Timothy Dennee, and Brian Crane. "The Receptacles Were Emptied of Their Contents": Archaeological Testing of Area II-B of the Carlyle Property and Excavation of the Shuter's Hill Brewery Site (44AX35), Alexandria Virginia." Parsons Engineering Science, Inc. Fairfax, Virginia, 1996.

Information on Shuter's Hill Brewery near Alexandria, Virginia.

Weeks, Morris, Jr. *Beer and Brewing in America*. New York: United States Brewers Foundation, 1949.

A short book covering the industry in the late 1940s.

Weisgerber Papers. Special Collections. University of Idaho Library, Moscow, Idaho.

This collection contains photos, deeds, and more about the brewery that was the first in the state and one of the longest lasting and largest in the area.

West, Elliot. *The Saloon on the Rocky Mountain Mining Frontier*. Lincoln: University of Nebraska Press, 1979.

A high-quality, scholarly study of the saloon that considers much of the relationship between beer and saloons.

Western Brewer 33, no. 3 (March 15, 1908).

Western Brewer 37, no. 6 (June 1912).

This was a journal for the brewing trade. It often failed to list authors or titles for proper credit.

Wilkie, Laurie A. "Documentary Archaeology," in *The Cambridge Companion to Historical Archaeology,* Dan Hicks and Mary C. Beaudry, eds. Cambridge: Cambridge University Press, 2006, 13–33.

Wilson, R. G., and T. R. Gourvish. *The Dynamics of the International Brewing Industry Since 1800*. New York and London: Routledge, 1998.

A study of the brewing industry in various nations for this time period. It provides background from a business perspective. Two chapters focus on the United States.

Wood, Donald F. *American Beer Trucks*. Osceola, Wisconsin: MBI Publishing Company, 1999.

Well-illustrated study of the transportation used by brewers and its development through the years.

Wright, Larry, ed. *The Beer Can*. Matteson, Illinois: Great Lakes Living Press, 1976.

Study of beer can history as the can collecting craze was reaching its height.

Yenn, Bill. *The American Brewery: From Colonial Evolution to Microbrew Revolution*. St. Paul, Minn.: MBI, 2003.

A good nonscholarly history of U. S. brewing.

Zanjani, Sally. *Glory Days in Goldfield, Nevada*. Reno: University of Nevada Press, 2002.

Discusses the brewery in Goldfield.

INTERNET SITES

"Central Park 19th Century brewery complex, Grand Forks, North Dakota." *Archaeo-Physics Shallow Subsurface Geophysical Survey,* Nov 28, 2010.

http://www.archaeophysics.com/central/

Information on the Dobmeier Brewery of North Dakota.

Comer, Elizabeth Anderson. "Public Archaeology as a Tool for Community Preservation and Empowerment." *Archaeology* Nov. 28, 2010. http://eacarchaeology.com/publicBalto.html

Information about Baltimore brewery archaeology.

Eurekalert-Science News Service. http://www.eurekalert.org/pub_releases/2004-12/uop-9ho120204.php

Patrick McGovern's work on finding residue of 9,000 year-old Chinese fermented beverages on pot shards.

From Site to Story—The Institute for Minnesota Archaeology website.

http://www.fromsitetostory.org/sources/papers/mnarch48/48inv-bf-a.asp

Contains specific, detailed accounts of Minneapolis brewery archaeology.

Montana Department of Commerce —Montana Heritage Commission website

www.montanaheritagecommission.com/reportsupdates/presentationcrewupdates/prescrewreppmay-aug06.pdf_

Offers information on the Gilbert Brewery in Virginia City, Montana.

"Portner's Brewery." Alexandria Archaeology Museum, Preservation in Action, City of Alexandria, Virginia, Nov. 28, 2010.

http://alexandriava.gov/uploadedFiles/historic/info/archaeology/ARSiteReportPortnerBrewingCoAX196.pdf

Information on Portner's Brewery near Alexandria, Virginia.

"Real Beer Page Mail, The Free Monthly Beer News Digest for the Online Beer Enthusiast." Subscriptions available at rbpmail@realbeer.com

Various issues, various dates, on recent brewery news.

Scalley, Shirley. "Shuter's Hill Brewery." Fairfax, Virginia: Engineering Science, Inc., 1996. Alexandria Archaeology Museum, Preservation in Action, no.35. Nov. 28, 2010.

http://alexandriava.gov/uploadedFiles/historic/info/archaeology/ARSummaryShutersHillBreweryAX35.pdf

Information on Shuter's Hill Brewery near Alexandria, Virginia.

INDEX

Page numbers appearing in *italics* refer to illustrative material.

ABOUT THE AUTHOR

Herman Wiley Ronnenberg was a teacher and librarian/media specialist at the elementary, secondary, and college level for thirty-eight years in California, Idaho, and Washington. He earned a Ph.D. in history with an emphasis in historical archaeology at the University of Idaho, and has worked on archaeological projects in Idaho, California, Arizona, Alaska, Virginia, Belize, and Peru. He is currently an adjunct professor of history at the University of Idaho in Moscow, Idaho, and the Institute for Extended Learning in Pullman, Washington.

Ronnenberg actively collects breweriana, specializing in artifacts from Idaho, allows public viewing of his collection, and belongs to all the major collector clubs. He wrote *Beer and Brewing in the Inland Northwest* (University of Idaho Press, 1993), *The Beer Baron of Boise, The Life of John Lemp* (Heritage Witness Reflections Publishing, 2009), and more than sixty periodical articles on brewing history. He lives in Troy, Idaho, with his dog, Shadow, and frequently visits his four grandchildren.